BROKEN
HEARTLAND

BROKEN HEARTLAND

The Rise of

America's

Rural Ghetto

AN EXPANDED EDITION

Osha Gray Davidson

University of Iowa Press ꕕ Iowa City

University of Iowa Press, Iowa City 52242
Copyright © 1996 by Osha Gray Davidson
All rights reserved
Printed in the United States of America

Printed on acid-free paper

Library of Congress Cataloging-in-Publication Data
Davidson, Osha Gray.
Broken heartland: the rise of America's rural ghetto /
by Osha Gray Davidson.–Expanded ed.
p.   cm.
An expanded ed. of the 1990 Free Press ed.
With an additional chapter.
Includes bibliographical references and index.
ISBN 0-87745-554-6 (pbk.)
1. United States–Rural conditions.   2. Rural poor–
United States.   3. Farmers–United States–Social
conditions.   4. United States–Social conditions–1980–   .
I. Title.
HN59.2.D35   1996
307.3'366'0973–dc20   96-12117
                                             CIP

04   05   06   P   6   5   4

Originally published in 1990 by the Free Press,
a division of Macmillan, Inc.

For Mary

My love, my life;
My dove, my wife.

Contents

Preface

America discovered the rural crisis in much the same manner as the New World was itself discovered: suddenly and by accident. One day American agriculture was riding high, widely hailed as the most technologically advanced and by far the most productive agricultural system in the world. The next day the entire farm population was in dire straits. Or so it seemed. Once-proud farmers—those larger-than-life, nearly mythic figures who inhabit our nation's collective unconscious, forever sailing combines as big as ships across seas of ripened grain—suddenly showed up on the six o'clock news, small and beaten down, choking back tears as the family farm was sold at auction.

The crisis that grips the Heartland has something else in common with America: the dimensions of both were at first vastly underestimated. The New World was originally thought to be a series of medium-sized islands. Who guessed that beyond the Caribbean archipelago lay two giant continents? Today, we are still sailing along the edges of the rural crisis; it is only the shoreline—what we call the farm crisis—that we know, and even that incompletely.

Part of the reason for this inability to see the larger picture is that we are a nation of specialists, each of us seeing the world from our own narrow discipline. What is a "shakeout" to an agricultural economist is a mass migration to a demographer, a changing market to an advertising executive, an overcrowded caseload to a social worker, a gender issue to feminists, a test of faith to the religious, and a descent into hell for the families in question. And so we balkanize the problem. We have a farm crisis and a hunger crisis. And crises of poverty, depopulation, water quality, deindustrialization, homelessness, soil erosion, transportation, education, drugs, child abuse, hate groups, banking, health care, and on and on.

But what is destroying rural communities is no more a farm crisis than the Boston Tea Party was the result of a tax crisis. The troubles in America's Heartland are symptoms of much larger problems in our society. Unless and until we confront these larger problems, future generations will be condemned to endure lives stained by poverty in ghettos, rural and urban.

Two final notes:

Some may argue that the lack of consideration in this book of the problems of American Indians is a serious omission. I would agree. It is serious because American Indians as a group comprise the most impoverished segment of rural people. Yet I would also argue that such an omission is necessary, since conditions on the reservations—as well as the historical circumstances under which reservations were formed—are unique and cannot be adequately addressed in a general book on rural problems.

Others may object to using the word "ghetto" to characterize rural communities. The term, after all, has roots in ethnic segregation, a condition not generally present in a rural context. But the word "ghetto" is, I believe, far more helpful in characterizing and analyzing the state of today's rural communities than it is misleading. It is my hope that the evidence contained in the following pages bears out that claim.

Acknowledgments

I am first and foremost indebted to the residents of Mechanics-ville, Iowa, for tolerating my prying presence in their town from 1986 to 1989. Special thanks to Jane Pini, who provided my introduction to Mechanicsville, and to Jo Randolph, proprietor of Doc and Jo's cafe, for gently nudging this project toward completion with her habitual greeting: "So when're we going to see that book?"

I am also indebted to the dozens of rural individuals who graciously allowed me to witness and record their oftentimes painful quotidian dramas.

Many people played a role in the development of this book. From the project's inception—and at many crucial points along the way—Daniel Levitas (first as research director for Prairiefire and then as executive director of the Center for Democratic Renewal) patiently explained the minutia of agriculture policy and the equally labyrinthine world of far right groups.

The entire staff of Prairiefire has my gratitude for answering countless questions about farm and rural life, and for providing me with key contacts in rural communities throughout the Midwest. I owe a special debt to Prairiefire's director, David Ostendorf.

I am equally indebted to Professor Michael Jacobsen for introducing me to the concept of rural ghettoization, as well as for his continued support, criticism, and encouragement along the way.

Dixon Terry, farmer and farm activist, was of great help, as both a knowledgeable source for and astute critic of my early articles on farm policy. He was killed while bailing hay on his Iowa farm in June 1989. I will miss his wise and compassionate voice.

A number of individuals—each expert in his or her field—generously gave their time to this project. They include Beverly Hannon, David Osterberg, Walter Goldschmidt, Mary Bruns, David Swenson, Bill Gillette, Joanne Dvorak, Deborah Fink, and Marian Meyers.

Special thanks to the staff of the government documents section at the University of Iowa Library for their hours of help in tracking down information, and to Mary Bennett of the State Historical Society of Iowa Library for her assistance in locating nineteenth-century primary documents.

Elsa Dixler, now literary editor of *The Nation,* furthered this project in several ways—acting as editor for the original article on which *Broken Heartland* is based, supporting my application for a grant from the Ford Foundation, and introducing me to my agent Gerard McCauley, whom I also thank.

Warm thanks to Joyce Seltzer, my editor at The Free Press, for not being satisfied with pleasant-sounding generalities.

The following organizations and agencies generously provided a wealth of information: Physician Task Force on Hunger in America; Council of State Community Affairs Agencies; Natural Resources Defense Council; Public Voice for Food and Health Policy; National Mental Health Association; Center for the Study of the Recent History of the United States, University of Iowa; Center for Applied Urban Research, University of Nebraska at Omaha; Conservation Foundation; Center for Democratic Renewal; National Rural Health Association; Northwest Area Foundation; Institute for Research on Poverty, University of Wisconsin–Madison; Domestic Violence Alternatives; Theology of Land Conference, Saint John's University; National Research Council, National Academy of Sciences; Office of Technology Assessment; Center on Budget and Policy Priorities; Iowa Farm Bureau; California Action Network; Center for Rural Affairs; International Ladies Garment Workers Union; Land Stewardship Project; Cooperative Extension Service, Iowa State University; Rural Coalition. Thanks also to the offices of United States Senators Tom Harkin and Charles Grassley and then-Congressman Cooper Evans for their help in obtaining government studies.

Without a grant from the Rural Poverty and Resources Program of the Ford Foundation, this book would not have been written. I thank the people there—and the program director Norman Collins in particular—for having faith in this free-lancer and for their dedication to the cause of furthering true rural development. The conclusions reached in this book are mine alone, however; they should not be taken as the views of the Ford Foundation. Thanks, too, to author Richard Critchfield for pointing me in the direction of the Ford Foundation.

To my daughter Sarah, who bravely agreed to move to a small town so that I could pursue this project and who then had the courtesy to pretend to listen to long explanations of U.S. farm policy over the next several years, I am eternally grateful. Though living at a greater distance, my stepdaughter Sienna has been equally understanding and every bit as gracious in making allowances for her distracted stepfather. To you too, Sienna, I am forever grateful.

I am indebted also to my parents, Penny and Sol Davidson, for their support, material and emotional, over the years.

Finally, to my wife Mary—who devoted her sharp editorial skills to eradicating from my writing the gratuitous adjective and the awkward phrase—my appreciation and love.

While many people aided and supported the production of this book, I alone am responsible for any errors it may contain.

Unfootnoted quotations are from interviews.

BROKEN
HEARTLAND

1

Decline and Denial

It is a fundamental illusion of American culture: the persistent celebration of rural life in the midst of its destruction.

HARLAND PADFIELD
in *The Dying Community*

Mechanicsville, Iowa

This handsome town is much like any of the thousands of rural communities dotting the gently rolling hills of the Midwestern prairie. Built along a narrow mile-long ridge rising out of open land in eastern Iowa, the town of slightly over 1,000 residents stands above unbroken fields of corn and soybeans like a ship at sea. The old-fashioned water tower at the center of town together with the augers and grain silos clustered around the nearby Farm Service Center reach into the Midwestern sky like masts and rigging. The fields surrounding Mechanicsville do, in fact, resemble an ocean—especially in the summer when the wind blows hard from the south, stirring the corn into waves that race to the horizon.

While there is no mistaking the fact that Mechanicsville is essentially a farm community, with its roots deep into the land, few people living within the town limits actually farm. Residents work at a variety of jobs, mostly low-skill, low-wage jobs in Mechanicsville or in one of the surrounding communities.

At 10:30 on a chilly Monday morning in spring, the town's business district, a two-block area of turn-of-the-century red brick buildings, is nearly deserted. A solitary car sits outside the post office, its motor running while the owner is inside picking up his mail. Across the street, a trio of beat-up pickup trucks are parked in front of the Village Inn, the downtown's one remaining cafe. Inside, four elderly men in seed-corn hats play pinochle at the bar while another group of men sit around a table drinking black coffee and telling me about their town.

Jim Cook, owner of the local hardware store, is the obvious leader of this last group. Cook is a World War II veteran with a shaggy mane of gray hair, a salt-and-pepper pencil-thin mustache, and a handshake you're not meant to forget. He is a die-hard conservative, a supporter of President Ronald Reagan from back when Reagan was still governor of California—that is, when it really meant something to be a Reagan man.

Over the next two hours Cook dominates the discussion, talking up the town's school system, volunteer fire department, and the principles that made Mechanicsville great: hard work, thrift, simple living, and, most of all, community pride.

"We want to show everybody else that we can do it better than they can," he says with a smile that shows he has no doubts about Mechanicsville's ability to always come out on top.

In the same firm tones, Cook ticks off the major evils of the day: greedy farmers, back-stabbing politicians, welfare mothers who "keep right on having kids," and people who don't shop locally. The two other men at the table exchange a quick glance over their coffee cups when Cook mentions this last ill. The issue of "buying local" is one of the town's sore points, especially with Cook, a topic that can escalate from hard words to threats of a fist fight in seconds.

The issue surfaced recently after Cook asked a local farmer why he made all his large hardware purchases 30 miles away in the city of Cedar Rapids instead of buying local at Cook's store.

"I'm really sorry," the man told Cook, "but they sell cheaper over there. I can't afford to shop at your store."

Cook said nothing. He just stared the man down and walked away. A week later the same farmer came into Cook's hardware store to buy two bolts—a purchase of about a dollar.

"Sorry," Cook told the man without a smile when the farmer laid the bolts on the counter at the cash register. "You'll have to drive over to Cedar Rapids for them. You can't buy them here."

The farmer thought Cook was joking. He wasn't. The man left the store threatening to pop Cook one in the nose and later sent his son in to buy the bolts.

" 'Course, I wouldn't sell them to him either," says Cook mildly, and takes a sip of coffee.

Cook demands even more of himself. He once wanted to buy a bed but found nothing at the local furniture store that quite suited him, so he drove over to Cedar Rapids, found the bed he wanted, and went back to order it locally. The owner of the store said he wasn't interested in ordering anything other than what he had in stock. Cook tried everything he could think of to get the man to order the bed, but he wouldn't do it. But Jim Cook does not give in that easily. He called the company that distributed the bed, pretending to be the owner of the furniture store across the street from his own business. When the bed was delivered there, Cook went over and paid for it, including a healthy retail mark-up.

Throughout our conversation, Cook jumps at any suggestion that his town is anything other than a vital, thriving village. When one of the other men at the table recalls the time when every building in town had a business going and observes that "it was a very prosperous place back then," Cook leans forward in his chair and directs his words with cold precision at the man who has just spoken: "It is still."

The other man, an ex-farmer now in his seventies, blushes and fumbles for words. "Oh, sure . . . that's right. She still is. She's a prosperous place."

The most that Cook will allow is that there have been some problems lately.

"Sure, we may be seeing some troubles due to the downturn in the farm economy," he says, "but nothing different from anywhere else. Look, we've been through this before, during the Great Depression. We've been here for 150 years and we've always gotten through. And we'll get through now."

With that, the interview ends. Cook has to get back to his store, and I have other interviews. We shake hands and I promise to stop by his hardware store on my way out of town.

A few hours later, interviews completed, I go to say good-bye to Cook. His store turns out to be a combination hardware–kitchen appliance–gun shop. Back behind the crock pots and hammers is a long glass case packed with guns and ammunition. Blue steel pistols and chrome-plated six-shooters lie behind the glass; dozens of rifles line the walls. The store is modern on the inside, which is surprising because of the building's saloon-style brick and wood front, the kind you rarely see except in Hollywood westerns.

"Oh, I remodel every now and then," explains Cook. "You have to if you want to stay in business for 110 years like we have. My Dad bought the place in the summer of 1926, but it's been a hardware store since 1876. 'Course, a lot of new places in cities are remodeling to make them look like they're 110 years old. That's 'in' now. But I get sick and tired of that old-fashioned look. You've got to keep moving forward."

We're standing by the cash register still talking when an old man walks slowly in, nods to Cook, and heads for the greeting-card section.

"You should talk to Everett Ferguson," Cook says, nodding in the direction of the old man. "He's 90-something, still lives in the house he was born in. Hey, Everett," Cook calls out. "Talk to this guy. He wants to know about Mechanicsville."

I walk over and introduce myself. Ferguson is a small man, dressed in a plain green shirt that is buttoned to the top and a gray sports coat. His narrow face is surprisingly smooth, as if he has outlasted even his wrinkles, but his hands look as if they were made of wax paper that had been crumpled into a ball and then smoothed out, leaving a fine network of sharp creases.

"Everett," calls Cook, "tell him what's happening to Mechan-icsville."

Ferguson doesn't say anything for some time. He stares at me through thick-lensed glasses that make his eyes appear large and liquid. I begin to wonder if he heard Cook, or, if he did, if he's capable of answering coherently.

Finally, just as I've decided that Everett Ferguson is lost in the mists of age, he answers in a voice that is slow and surpris-ingly deep. "What's happening to Mechanicsville?" he asks with a scorn reserved for those who ask the obvious. "It's dying."

The words seem to hang in the air. I hear Cook suck in his breath as if about to say something, and I turn to face him. Cook is standing silently at the cash register, one hand on the counter, the other in his sweater pocket. He looks out the store's large front window to where the late-afternoon sunlight cascades down the facade of an empty building across the street. His face is empty, too; suddenly gone is the mask of bel-ligerent optimism, replaced by a new face—or a new mask—this one of studied indifference. It is as if he hadn't heard Fer-guson, as if he were alone, waiting out the last few minutes un-til closing time in the store his father bought back in the sum-mer of 1926.

The similarity of the prairie to an ocean (which is something of a paradox, since the Midwest is about as far as you can get from an ocean in this country) was noted immediately by the earliest

white explorers and settlers who christened the region the Inland
Sea. Judging from their diaries and letters, it was not so much the
wave-like motion of the prairie grasses that inspired the name, but
rather the emotions stirred in the settlers when confronted by
something so vast it hinted at the infinite.[1]

"I had the feeling that the world was left behind, that we had
got over the edge of it, and were outside man's jurisdiction,"
wrote novelist Willa Cather. "This was the complete dome of
heaven, all there was. Between that earth and that sky I felt erased,
blotted out. . . . That is happiness, to be dissolved into something
complete and great."

Many of the early Europeans didn't find the experience of dissolution quite so idyllic. Some, in fact, were terrified by the dimensions of the open land—a quarter of a billion acres of
shimmering chest-high grasses stretching from Illinois west to
what is now Kansas, and from the Dakotas south through Oklahoma and into Texas.

"Wherever a man stands he is surrounded by the sky," wrote
the stunned diarist of the conquistador Coronado in 1541, as the
party of Spanish explorers huddled around the campfire on the
Kansas prairie. Bewildered by the scale of the land, the group
stayed only long enough to satisfy themselves that there were no
"cities of gold" to be found and then hurried back to New Spain,
where the vistas were more manageable.

Mechanicsville's first residents were neither as enamored by the
prairie as Cather nor as frightened by it as Coronado's group.
They were a hard-headed, pragmatic, nose-to-the-grindstone con-
glomeration of German, Scandinavian, and Scotch-Irish pioneers
who drifted into the territory west of the Mississippi River in the
mid-1800s. They came from both the Yankee East and the deep
South to form a new society of farmers and shopkeepers whose
values, culture, and even dialect showed the influence of the two
strains.

The Southerners brought with them a high regard for gener-
osity and liberty combined with an almost visceral distrust of authority. That last trait has always been particularly strong in this
area. In 1931, when the government began testing all dairy cows
in Iowa for tuberculosis, scores of armed area farmers vowed to
shoot the first son-of-a-bitch to touch a Cedar County cow. The
National Guard had to be called in to protect the veterinarians.[2]

"Not that we thought it was a bad idea to test for TB," recalled a local farmer who was a teenager during the Cow War. "In fact, most everybody thought it was a good idea. We just didn't like being told we had to do it."

Mechanicsville's residents could be as ornery with private officials as they were with public authorities. The town once had a railroad depot on the south edge of town that serviced 12 trains a day—one every two hours around the clock. In November of 1867, a spark from a passing train landed on the wooden-shingled depot roof, setting it on fire. Because many townspeople felt the railroad hadn't been very helpful hauling firewood a few winters before, they decided to pay the railroad back. People rushed down to the depot and instead of helping to put the fire out, they stood happily by, watching the building burn to the ground.[3]

But these attributes have always been tempered by a Southern respect for hospitality and good manners—attributes that are characterized in Iowans by a tendency to politeness that often borders on the absurd. When 1988 Democratic vice-presidential candidate Lloyd Bentsen received a few scattered boos from the crowd at the Iowa State Fair, a campaign spokesperson noticed that the catcalls were more subdued than at other stops. "In Iowa, even the hecklers are pretty polite," he observed.

The Yankees added respect for education, dedication to hard work, and a stern puritanical morality. A foreign visitor dubbed the resulting Midwestern amalgam "the most American part of America."[4] That observation was echoed during the 1988 presidential caucus when a visiting Italian journalist called the small-town Iowans he encountered "the original Americans, as if preserved in amber."

A visitor traveling through Mechanicsville today would probably agree with that assessment. From the straight tree-shaded streets, with their large old houses and sprawling wrap-around porches, down to the neatly trimmed front lawns edged with rows of petunias, the town looks as if it belongs to an earlier era. For generations, Mechanicsville has remained the knot tying together the lives of the farm families who till the rich black earth in this small piece of America's Heartland. They went to school here and shopped in the modest downtown. They socialized here, attending dances at the American Legion Hall and softball games at the dusty field by the railroad tracks. They were married here at one of the three churches (Catholic, Presbyterian, and Methodist), and on anniversaries they feasted on steak and potatoes at the restaurant Our Place. When the nearby farm couples grew old, they passed their farms down to their children and moved to town. And when at last they died, usually at home among family and friends, they returned to the earth here, buried beneath the prairie grasses in the Rose Hill Cemetery on the west edge of town.

Like most small towns, Mechanicsville always had trouble holding onto its young. Many felt stifled here, their possibilities too limited, the pace of life too slow. And so every year one or two of these ambitious young men and women left for the bright lights of Des Moines or the even brighter ones of Chicago, Minneapolis, St. Louis, or beyond. But many remained, settled into small-town life, and raised families. In fact, over one-quarter of Mechanicsville's residents have lived in the area for more than forty years; nearly 80% have lived there for at least a decade.[5]

Jim Cook was one of those who stayed. "When I got out of the service," he recalls, "people said, 'What the hell you come back here for?' I looked them right in the eye and said, 'I've been every place I could be, and Mechanicsville is no different from the rest of them. It's just as good as Carmel, California, or Timbuktu. Every one of them has their faults and if I'm going to have faults, it's going to be with the people I know. That's why I came home: because it's home.' "

That longing for a community that is "home," the need to feel part of a group that is larger than a family but more embraceable than a nation, is a familiar theme throughout American social history. Since the earliest days of settlement, rural communities have satisfied that desire by playing a wide variety of roles. "It is the community that cushions pain, the community that provides a context for intimacy, the community that represents morality and serves as the repository for old traditions," observes sociologist Kai Erikson.[6]

Life in the tightly knit rural community of Mechanicsville has always been profoundly different from that found just thirty miles away in Cedar Rapids, with a population of around 100,000. The main difference between the two is that while most Mechanicsville residents have always been essentially united—whatever factors happened to divide them—in Cedar Rapids, residents have always been essentially divided—whatever factors happened to unite them.[7]

Eleanor Anstey, a professor of social work at the University of

Iowa, recalls an incident from her high school days on an Iowa farm that for her sums up this experience of life in a rural community: "I telephoned the local flower store for lilies, but they said they were sold out. Suddenly, a voice on the party line said, 'Oh, I've got some nice ones you can have, Eleanor.' It wouldn't have occurred to you to feel that your privacy was violated."

Of course, it is easy to idealize small towns such as Mechanicsville, to forget the schisms, economic and social, that *do* exist there. It's easy, too, to ignore the currents of racism and anti-Semitism that run just below the surface, currents that appear in crude but relatively harmless jokes—or in far uglier ways in hard times. And one needn't be African-American or Jewish to feel shunned in a small town; the Yankee inheritance of puritanism allows for little deviation of any kind. For example, a woman who chooses to pursue a career while her husband stays home to raise their children can expect little support from the community for such a decision.

Besides, not everyone appreciates the kind of intimacy a small town provides. A 1981 survey revealed that almost half of Mechanicsville residents felt their neighbors interfered in their business too often. But 90% of respondents also believed their neighbors would help out in an emergency, and for most, the trade-off was worth it.[8] For all the drawbacks to small-town life, that sense of belonging to a caring community is what Heartland towns like Mechanicsville have always provided their residents.

But today that is changing. Small towns are in trouble. Strictly speaking, Mechanicsville and the thousands of rural communities like it are not dying, as Everett Ferguson put it. To use the term "dying" in this way at once overstates and understates the problem faced by small-town residents.

It overstates the problem, in literal terms, because most rural communities will survive—at least they will have residents and so will remain on the map for decades. But in many ways, the situation would be less dire if the towns simply folded up and the residents moved away. Instead, formerly healthy, mostly middle-class communities throughout the Midwest, the small towns that have given the area its distinctive character since its settlement, are being transformed into rural ghettos—pockets of poverty, unemployment, violence, and despair that are becoming more and more isolated from the rest of the country. As the coastal economies have boomed, the Heartland has collapsed. "The most

American part of America" is fast becoming "America's Third World."

The dimensions of the problem are sobering. Between 54 and 60 million rural Americans, one-quarter of the country's population, are touched by the decline. Over 9 million people now live in poverty in America's rural areas.[9] In Iowa, the hardest hit of all Midwestern states, one out of six individuals falls below the federal poverty line, and in some counties the poverty rate approaches 30%.[10] With an irony that is especially bitter in this region, the nation's breadbasket, hunger has become a common problem.

"We've seen a steady and continuing increase in the need for food in the past five years," says Karen Ford, director of Food Bank of Iowa, which supplies donated food to 200 food pantries and nonprofit agencies throughout the state.

As the economy stagnates, manufacturers lay off workers or shut down completely. Hospitals, banks, and businesses close. Depression, suicides, and child-abuse rates grow. The need for foster care rises to an unmanageable level as families break up under the pressure of poverty. Towns compete for factories paying poverty-level wages. Mass migrations become commonplace. Local governments cannot afford the most basic services.

"People talk about the middle class being in jeopardy in Iowa, but that's inaccurate," says University of Iowa economist David Swenson. "A significant portion of the state is already out of the middle class. The notion of upward mobility in Iowa is gone."

It is ironic that the victims of this blight, the inhabitants of the new rural ghettos, have always been the most blindly patriotic of Americans, the keepers of the American dream. Their response to any criticism of America was summed up in the bumper sticker that was once common around here: AMERICA, LOVE IT OR LEAVE IT. That patriotic decal can still be seen on pickup trucks throughout the Heartland, but today it competes with another bumper sticker that reads: SHIT HAPPENS.

The speed with which the recent decline hit rural America has made the problem even more difficult for Midwesterners to deal with. Iowans, especially rural Iowans, are well known for their resistance to quick changes of any kind. A retired farmer once told me that his father was the first person in the area to try raising soybeans back in the early part of this century, when corn was the undisputed king.

"It probably took quite a while to catch on," I remarked.

"Oh, no," he assured me. "Why, some of the neighbors were giving the new crop a try just six or seven years later."

And so the reaction to the decline over the last few years has been, as usual, to wait it out—to endure. But this catastrophe is not like a period of drought that can be outlasted. Whatever recoveries may temporarily come this way, short of major structural changes in our economy and government, the rural problem is here to stay. According to a study prepared for the U.S. Congress's Joint Economic Committee, "Iowa could become the State that the Nation leaves behind."[11]

Despite the magnitude of the problem, the disintegration of rural America is largely an invisible crisis. Driving along Interstate 80—the way most outsiders see the state—you would never guess anything is wrong. From that narrow corridor you drive for hours passing fields of corn and beans that cover the horizon in lines as straight as a table's edge. Giant tractors or combines crisscross the land, planting in the spring, cultivating or spraying in the summer, harvesting in the fall. Everything you see speaks of abundance and prosperity.

Even for those few adventuresome souls who pull off the interstate and head into small farm towns like Mechanicsville, appearances are deceiving. The disaster that is sweeping through the Midwest is not like a tornado or a flood that leaves a trail of rubble and twisted-up cars in its path. (For this reason the rural crisis makes for poor film footage and so doesn't rate a spot on the nightly news.)

But if you look carefully at downtown Mechanicsville, you will notice that although the buildings still stand, a majority of them stand empty. At one time the town had as many as thirty Main Street businesses. There were two feed stores, two farm implement dealers, two hotels, two clothing stores. Also a pharmacy, a jewelry store, a soda fountain, a shoe store, an opera house, a pool hall, a bakery, a butcher shop, and a produce market. Today none of these remain.

Kathy Lehrman and her husband Kelvin bought the local paper, the *Pioneer Herald*, in 1979. In the spring of 1986 she stood outside her downtown office looking across the street at a row of empty storefronts.

"I don't know what's going to happen here," she confided. "In the past six months we've lost ten businesses in the three towns we

cover. Maybe somebody ought to come in, buy the whole downtown, and just tear it all down."

One month later, the *Pioneer Herald* office was also dark. The
Lehrmans had sold the paper to a chain and were looking to try
their luck somewhere else.

"People are hoping for things to get better," says 49-year-old
Steve Seehusen, who runs a combination real estate firm and insurance agency from an office in what used to be a bank (until it
closed in the Great Depression). "These are hard times. We used
to sell houses as fast as they came on the market. Right now I've
got seven houses listed, and they're just sitting there."

Even with homes that sold for $45,000 just a few years ago now
selling for $25,000, there are no takers. "So many things are out
of our control," says Seehusen dolefully.

The creation of rural ghettos is a complex process, and despite
the rapid changes of the last decade it has been evolving over
several generations for reasons that are less than obvious. To
understand the decline in America's Heartland we have to start
with the well-known but little-understood event associated with it:
the farm crisis.

2

——

Roots of the Farm Crisis

Agriculture is not the problem. Agriculture is doing just fine. It is the people *who are having problems.*

CORNELIA BUTLER FLORA,
Rural Sociologist

Keokuk, Iowa

The lay minister stands with upraised hands before a crowd of farm families gathered under gray skies and offers up this prayer:

"Lord, help us to forgive anyone who has hurt us these past months—partners, family, neighbors, money lenders, or anyone else. We pray to the Lord."

"Lord, hear our prayer," answers the crowd.

"Lord, we come before you in our brokenness, longing for your gentle touch and loving embrace. Your wounded body and spirit heals us, your brokenness restores us to wholeness. Praise to you, our God, for your mercy and kindness. Amen."

"Amen," says the crowd. Together, they conclude the prayer service by singing "America, the Beautiful."

Five minutes later, at 10 a.m., just as a cold rain begins to fall, the sheriff's sale begins. Kathy and Mike Bolin's farm, like that of so many before and after them, is on the auction block.

Kathy, who had been fighting for over two years to avoid this day, stands beside her husband and four children as the few pieces of machinery they had accumulated are sold.

But she doesn't hear the chant of the auctioneer; she doesn't see the knot of men, their hands thrust deep into their overall

13

pockets as they stand around the Bolin's well-maintained old John Deere tractor. She is deep inside her own mind, wondering what to do with the body of a favorite dog that died earlier that day. The question at first tugs gently at a small corner of her attention—a detail that simply wants an answer. But soon the issue is all she can think about.

"Where do you bury your dog after you've lost your farm?" she wonders over and over. It is the first challenge of Kathy's new life off the farm.

When the last piece of equipment is gone and the crowd heads for home, Kathy feels nothing except the chill of the rain on her hair and the equally cold edge of the first unanswered question.

It is now another spring morning two years after the Bolins lost their farm at the sheriff's sale, and Kathy Bolin is no longer numb. She is angry. Of course, the anger doesn't show plainly—that isn't the way it's done in Iowa. But from time to time as she talks about the farm and how they lost it, the muscles of her neck tighten and her voices goes flat, and finally Kathy has to stop talking altogether and wait for the emotions to wash over her like a wave and finally recede.

The ex-Californian with short brown hair and gray-green eyes sits at her kitchen table stirring a cup of instant coffee. The walls are unpainted plasterboard. Behind her, a wood stove sits idle near the center of the room.

Outside it is a spectacular May day. Warm and bright and cloudless. The kind of morning that makes Iowans quickly forget the bitter Midwestern winter. From the kitchen window directly in front of her Kathy looks out onto the fields that she and Mike used to own. Someone else's tractor trundles back and forth across the land, preparing the soil for planting.

She is waiting for a phone call. In a deal worked out with their creditor, the Federal Land Bank, today is the last day for the Bolins to come up with the money to keep their farm house and a couple of acres of land around it. It's not enough land to farm, of course, but at least it would allow them to stay in the house and raise a few horses and a couple of cows. Mike is several hours late in coming up with the money, and Kathy is afraid that the lender will use the delay to force them out of the house, and so she waits by the phone.

There is bad blood between the Bolins and the bank. The foreclosure, which the family fought every step of the way, caused hard feelings all around. After losing the farm, Kathy wrote to the bank asking that she and Mike be allowed to rent the land back. Her former loan officer set up a meeting. When she showed up at the bank office, the loan officer walked up to her and calmly stated, "I don't have to rent to you. You people have been nothing but a pain in the butt."

"He could have told me no over the phone," says Kathy, her voice going flat again with suppressed anger. "He called me in just so he could humiliate me."

Things were different in 1979 when Kathy and Mike first picked out the land and applied for a loan, for it was then the peak of agriculture's boom years, a feverish period that began in 1972 when Secretary of Agriculture Earl Butz issued a clarion call to farmers to "get big or get out." The United States had embarked on an ambitious program to vastly expand the country's agricultural export trade.

As with all government policies, this one had a number of objectives, domestic and international. Election-year politicking played its part. President Richard Nixon saw the plan as a way to increase farm income and so capture the farm vote.[1] The drive to sell more U.S.-grown grains to other countries was also an attempt to shore up an increasingly unfavorable balance of trade, a problem caused by declines in foreign sales of U.S.-manufactured goods and increases in the price of oil.[2] And for all the rhetoric about "feeding the world" surrounding the new policies, government-subsidized food sales were granted or denied more often for geopolitical reasons than because of humanitarian concerns.[3]

But whatever the motives of its backers, the venture had a dramatic effect on the farm economy. The 1970s were a heady time of rapidly rising farm income and skyrocketing land values, and few questioned the notion that the good times would roll on forever. Banks urged farmers to take out larger and larger loans to modernize and to expand operations. One Iowa farm family applied for $12,000 in 1979 only to find their check made out for $25,000. When they called their loan officer about the mistake, he just laughed. "Don't be foolish," he chided them. "Go ahead and use the extra money for whatever you want. You're good for it."

But they weren't really. Like virtually all farmers in the area, unless they sold their land and quit farming they were flush only on paper. As inflation climbed in the seventies, farmland values soared across the country. In Iowa, the average price of an acre of farmland rose from $419 in 1970 to $2,066 in 1980, with no ceiling yet in sight.[4] For the average farmer, whose land comprised about three-fourths of his total assets, it was an unprecedented bonanza: his net worth in land jumped almost a half-million dollars during those years.[5]

"Why not put your assets to work?" became the rhetorical question of the day. And the best way to do that, according to virtually all the agricultural experts and bankers, was for farmers to borrow against their inflated holdings and get bigger. Millions of farmers followed that advice. They traded in their 55-horsepower tractors for 200-horsepower behemoths. Of course the price for these machines climbed right along with their engine size. While a prudent farmer could pay $10,000 for a decent tractor in the early 1960s, he or she would have to shell out at least $80,000 for a large model in the 1970s. But with these expensive machines farmers could also farm more acres. In fact, they needed to, to earn enough to finance the bigger equipment.

And all the while farmers continued to borrow against their still appreciating land. In 1971 farm debt stood at $54 billion. By 1985 that amount had swelled to $212 billion—a figure greater than the combined debt of Brazil, Mexico, and Argentina. American agriculture had become the most capital-intensive system of food production in the world.[6]

Like many young couples, the Bolins borrowed not to expand, but to buy their own land—in 1979 they were still renting land from Mike's father.

"It was a dream come true for Mike," says Kathy. "He was born and raised on a farm. It was what he always wanted to do."

The Bolins used their life savings and sold almost everything they owned to come up with the $56,000 deposit on the 160-acre farm. Like most of the farmers in the area, they planted corn and soybeans and raised a few cows. Mike also had another, less common passion: raising draft horses. He soon earned a name for himself in surrounding counties because of the excellent quality of his horses. Besides raising horses and farming, Mike also worked as a union electrician on construction sites around the area to bring in the money the family needed to keep up with farm payments.

It didn't take long for things to sour. While the U.S. government was still pushing policies to increase farm production, other countries were now producing record crops themselves, thanks to both technological and political changes. From 1970 to 1981 U.S. production of grain expanded by 20%.[7] At the same time, production in China, India, and Brazil (all three formerly large grain importers) also rose. The flooded markets naturally led to lower grain prices worldwide. At the same time, the government also lowered its support prices, in an effort to hold on to foreign markets by making U.S. grains cheaper.

Like a balloon loosing hot air, grain prices began to drop. Slowly at first, losing a penny or two per bushel at a time. And then faster. A circular process with devastating results for farmers began. In order to make up for diminishing profits, farmers were forced to grow more crops. The increased production forced grain prices even lower, which in turn led to farmers' putting more land into production, causing commodity prices to drop lower still.

But falling prices due to increased worldwide production and lower support prices weren't the only problems facing farmers. On October 6, 1979, the Federal Reserve raised the cost of borrowing money, a commodity that is as precious to farmers as rain. According to Iowa State University agricultural economist Neil Harl, the resulting rise in interest rates "threw agriculture into the windshield." At the same time, land values—which are closely tied to commodity prices—started to slide. The downturn in the farm economy affected other areas. Construction work, which the Bolins needed then more than ever, came to a virtual standstill.

While the interest rate on the Bolin's loan shot up from 8.75% in 1979 to 14.75% in 1983, the value of the state's farmland—the golden equity against which farmers like the Bolins had been borrowing—plunged 63% in just five years. Between 1982 and 1985 U.S. farmland values fell $146 billion, a massive figure equal to the combined assets of IBM, General Electric, Kodak, Procter & Gamble, Dow Chemical, McDonald's, RCA, Upjohn, Weyerhaeuser, and CBS.[8]

Average net farm income in Iowa went from $17,680 in 1981 to $7,376 in 1982. In 1983 the figure fell to $-1,891.[9]

"We're going from crisis to chaos out here," an Iowa minister told the press. The bubble had burst.

"Everything from that period is hazy," says Kathy. "I think that's from choice."

What happened in those dark days was repeated in thousands of farm families across the Midwest.

"In 1984 we were paying $5,000 to the [Federal Land Bank] every six months," says Kathy. "We had been behind before, but had always caught up. In March 1985 we paid $3,000 and the next week we got a notice of foreclosure. The sheriff came out. The kids got to know him real well around that time. He was nice about it, though. You could tell he didn't want to be doing this. He shouldn't have to. The loan officers should have to face us and the kids—they forget that there are real people involved in this mess.

"We were told that by July 1 we needed to pay them $14,500. Well, Mike was finally working a good construction job over in Illinois, bringing home $1,000 a week sometimes. The loan officer showed up at the house on July 1 wanting the money. I told him that we had $12,500 and could get the rest in a month. I explained about the new job Mike got. But he said no, we couldn't have another month. He said he wanted whatever we had now and that then he was going to foreclose on us. Well, I told him that if he were going to foreclose on us anyway, I wasn't going to give him the $12,500. He just laughed and said he'd follow me to the bank and get the money there."

Kathy sighs at the bitter memory of those days. But even harder times were to follow. The Bolins decided to fight foreclosure. As money grew tighter, they were forced to give up more and more. Luxuries like vacations and entertainment were the first to go. Then clothes, insurance, and any household goods. The Bolin children didn't even ask their parents for toys at Christmas. "They knew we couldn't afford any gifts," says Kathy. The phone was disconnected, increasing the family's sense of isolation. As money grew more difficult to come by, the specter of hunger became real. Friends stopped coming by. "Maybe they thought foreclosure was contagious," says Kathy with a grim laugh.

Under the strain, the family started to crack. Mike blamed Kathy for their problems and avoided the house. When he did come home, savage fights erupted over the smallest issues. By late summer 1985, Kathy found herself considering suicide.

"I just started not liking myself," she says. "I cried all the time. 'If only I had just loved this man more,' I thought. 'If only I had worked a little harder.' I felt like I had failed them. Mike. My kids. Everyone."

Then, in September, Kathy took the children and went to California to visit her parents. The real reason for the trip was to get away from a situation that was driving her to the brink of insanity. She considered staying on the coast. After all, she reasoned, why return to a losing battle? Her old employer heard she was in town and offered her old job back.

Just as things were coming together for Kathy in California, she decided to return to Iowa. The decision surprised everyone, including herself.

"I knew that we might not win this fight," she says. "We could lose the farm. But I also realized that at least if we stayed together and fought this thing, we'd always be a family. That much was up to us."

Kathy and the children drove home to a tearful reunion with Mike. The foreclosure notice arrived a few days later.

Just as Kathy reaches this point in her story, the phone rings and she jumps up to answer it.

"Yeah? What did they say?" She leans against the wall and listens for a few seconds, nodding silently. Then her face lights up and she places a hand to her throat. "They did? Well, Thank God. I'll see you later," she says, hangs up the phone, and sits back down at the table. "This is a milestone day," she says. For the first time that day, Kathy allows herself a small smile.

The door to the back porch suddenly opens with a loud creak and in comes the Bolin's eldest daughter, 12-year-old Annie, her boots echoing on the bare wood floor. She had been out in the barn tending a Belgian colt born just before dawn that same day. Her blonde hair is tied straight back from her face, and she is dressed in a red Iowa State T-shirt and mud-caked jeans. There is something reminiscent of a colt about Annie herself, an awkward, endearing quality evident in the shy way she steps across the kitchen, stops by the stove, and glances up at the wall clock.

"I hope Dad calls soon," she says.

Her mother turns in her chair. "He just did. We can stay."

"Oh goody, goody," Annie squeals, and then blushes and stuffs her hands into her jeans pockets, embarrassed at her un-Midwestern outburst of emotion. She quickly exits, out onto the back porch, where she stands for a minute looking out toward the horizon, and then races off to check on her colt.

"Well, I guess now I'll unpack the curtains," says Kathy, eyeing a cardboard box that sits beneath a bare window in the living room. They have been there for over two years, packed away when the Bolins thought they'd have to move. "Didn't seem like any point in putting them back up until we knew we could stay," she says. It's then that I notice the many boxes of all sizes stacked in every room and realize their significance: the family has been living in limbo for over two years.

"I certainly don't feel vindicated," she says. Her joy at winning the battle cannot match her bitterness over losing the war. "We got a raw deal. What about the farm? When we first moved in almost ten years ago, we planted an orchard: apples, cherries, a lot of different fruit trees. The first thing the people who rented the fields did was to bulldoze those trees down—so they could get a couple of extra rows of corn in. You can't describe the kind of pain you feel watching that. And then they plowed up a rhubarb patch out back that was planted in 1901. It makes you sick to see that.

"We were going to pass this farm down to our children, see? So we took care of everything. It wasn't just 'Let's see how many bushels we can get off of this piece of dirt.' Mike would actually disk around a quail's nest in the field. He's the Jacques Cousteau of Lee County." She laughs at the thought and then looks back out the window. "It's real hard to watch them farm," she says, shaking her head. "They don't give a shit about why the land is here. The first time the new farmers came, Mike told them to get off the land or he'd blow them away. Oh, God, he'd have done it, too. He thinks heaven is here on earth, on this farm.

"Sometimes I still get so down, and Mike will tease me. 'Come on,' he'll say, 'why don't you smile a little just to let me know you're glad to be alive?' " Tears fill her eyes. "Really, I couldn't have picked a better man," she whispers.

Kathy looks up at the clock and brushes a few invisible crumbs off the table. She has an hour before she must leave for work at the local hospital, where she puts in 25 to 30 hours a week. She is earning her nursing degree at the community college and plans on working at a hospice someday. There is a grim bit of logic to it—going from family farming to caring for the terminally ill.

"You never really get over this," Kathy says as she stands up. "And it's been rough on the kids. At school a while back, at show-and-tell, our youngest was telling about a new colt we had. 'You don't have a new colt,' said a little girl. 'Yes we do,' our boy said. 'No, you couldn't have,' the girl said in font of the whole class. 'You're too poor.' "

Annie clomps back in, grinning from ear to ear. "Want to see my horse?" she asks, and leads us out to the corral where a colt the color of taffy with a white zigzag on its muzzle stands wobbly-legged, nursing from its mother. I tell her that the colt is beautiful and Annie blushes a deep crimson, just like a proud mother being complimented on her newborn baby. She walks off to the barn to get a curry comb. Kathy watches her daughter go.

"I think the worst part is that there is no safe place anymore," she says. "Everybody should have one safe place. But at any time of day people can come in and invade your life for $2,000. Or doze down the trees you planted and took care of for years. Will my kids ever really feel safe after this? I don't know."

Kathy absently strokes the flanks of the big Belgian mare. The horse closes its long-lashed eyes and gives a sigh that comes from deep down in its massive body.

"I used to play a game when I was a little girl," Kathy says. "I would close my eyes and pretend that I went to heaven. God would be sitting there on a big chair. I'd climb up on God's lap and I'd tell him all my problems and he'd listen. And I'd feel better. I find myself doing that again. Only I can't tell him my problems. I'm scared to. What if my prayers aren't answered? So rather than take that chance, I don't say anything. I just sit on God's lap and I don't say a thing."

Kathy stops patting the horse.

"You can still pursue your dreams in America," she says, staring at the ground. "You just can't obtain them."

Despite the charges leveled by many liberal politicians, the farm crisis of the 1980s did not spring fully formed from the policies of Ronald Reagan. The Bolins were only the latest victims in a complex struggle that began more than a century before the affable president was born. It's for this reason that some close to the issue object to the term "farm crisis," with its implication of a sudden occurrence. They prefer to talk about the "farm condition." There is a song which bears out this idea:

So it goes, the same old story, with the farmer as a goat.
He can only pay his taxes and the interest on his note.
Oh, it's fun to be a farmer and to till the dusty soil,
But the guys who farm the farmers are the ones who get the spoil.

While it is easy to imagine Willie Nelson singing that song at the Farm Aid Concert in 1986, it was already popular with Heartland farmers in 1886. In fact, the battle lines were clearly drawn by then and the farmers were shaking the rafters of meeting halls throughout Nebraska with the anthem and cussing out the bankers and railroad companies—two of the "guys who farmed the farmers." Indeed, the great American struggle between the consolidation of power and great wealth for the few and a distribution of these commodities among the many—the battle, in effect, over the ability of the average American to achieve, not just pursue, the American dream—began even as the country itself was being founded.

At first the fight centered on issues of land ownership—not surprising in a society that was still mostly agrarian in the eighteenth century. To a people whose ancestors had only recently been peasants in Europe, owning land was seen not only as a path to wealth but also, more importantly, as the sole guarantee of freedom. As one colonial farmer wrote, "We have no princes, for whom we toil, starve, and bleed: we are the most perfect society now existing in the world. Here man is free as he ought to be. . . ."[10]

Among the American revolutionaries, Thomas Jefferson was perhaps the most passionate advocate of a government policy to distribute land to the common people, praising the small landholders as "the most precious part of a state," and "the chosen people of God."[11] Jefferson held these "yeoman farmers" in esteem for a variety of reasons, some of which sound quite crude to our twentieth-century ears. Although he later modified his stand, Jefferson in his early years railed against all aspects of urban life, writing that "the mobs of great cities add just so much to the support of pure government, as sores do to the strength of the human body."[12] Far better, he wrote, was the pastoral life, which produced in its adherents a variety of virtues, including diligence, industriousness, physical strength, and honesty.

But for all his questionable notions about the inherent superiority of rural life, Jefferson's allegiance to the yeoman farmer was based not so much on simple-minded agrarianism as it was on his belief that family farmers would form the cornerstone of a strong and lasting American democracy because they (1) were economically and politically independent from a ruling elite, and (2) had a stake in the fortunes of the new country because they owned land. From Jefferson's day to the present, behind most appeals to "save the family farm" lie claims to these twin virtues of freedom and entrepreneurialism.

To create this nation of yeomen, Jefferson advocated dividing up the country's vast fertile lands into modest parcels and turning them over to small farmers at no charge. To maintain this society of equals, he advocated a progressive tax on estates. But few eighteenth-century leaders were willing to go as far as Jefferson in creating specific government policies to ensure political—and especially economic—democracy. Alexander Hamilton, for example, sniffed at the very idea of democracy. "The people," he insisted, "are turbulent and changing; they seldom judge or determine right."[13] Rather than trust the common people—whom Hamilton sometimes referred to as "the great beast"[14]—with the important and difficult business of governance, Hamilton favored rule by "the rich and well-born." He supported the idea of a president and a Senate appointed, for life terms, from the upper crust. The land policy Hamilton advocated reflected this bias: he wanted federal land to be sold freely to the highest bidder, a policy which would benefit the federal treasury and land speculators the most.

In Jefferson's opinion, Hamilton's free-market approach was antithetical to democracy, for it would simply replace the Tory privileged class with an indigenous one. Wealth and power would eventually concentrate in the hands of a few large landholders, small farmers would be squeezed out completely, and democracy would wither.

These questions provoked bitter arguments between Secretary of the Treasury Hamilton and Secretary of State Jefferson. During cabinet meetings, Jefferson wrote, the two often clashed like fighting cocks.[15] It is hard to say, even today, who won that war. The larger battle was comprised of many smaller skirmishes, and both Hamilton and Jefferson won their share of those. Jefferson, for example, was midwife to the birth of a true participatory democracy; there would be no King George Washington, nor a Senate of bluebloods elected for life.

Hamilton, however, was far more successful in the economic realm. The nation's first land policy was largely the one he favored. Most federal land was sold on the open market to speculators who then resold the land to farmers, often at inflated prices. Largely because of this policy, Jefferson's dream of a true egalitarian society of independent farmers never materialized to the degree he wished, or, for that matter, to the degree commonly believed.[16] The East quickly became industrialized and urban; in the South, large plantations worked first by slaves and later by sharecroppers dominated agriculture. Government-sponsored irrigation projects which were allegedly designed to help small family farmers, but whose benefits were reaped chiefly by the already wealthy, gave rise to giant empires in the arid American West. Only in the Midwest, in the American Heartland, did the independent family farmer and small-town resident achieve a long-lasting reign. And even there the victories were usually adulterated and always tenuous. "Despite its place in the nation's value system, widespread private ownership was a myth by 1880," writes rural sociologist Charles Geisler.[17]

But how did this happen? After all, America was known from the start as the land of opportunity, the one country in the world where even the humblest individual could raise himself or herself up to a decent and honorable life. That, in a nutshell, *is* the American dream. But somewhere between the words "We, the people . . ." and the dotted line at the bottom of the land deed, something went very wrong.

The Homestead Act of 1862, under which Jefferson's dream of an agrarian nation of small landowners was finally to have been made a reality, has been called "the greatest democratic measure of all history." But despite some substantial successes, the Homestead Act never managed to create a nation of yeoman farmers.

The provisions of the act were straightforward: settlers were to be given clear title to 160 acres of farmland after they lived on the property for five years and made certain improvements to it. But the act was doomed from the start. By 1862, the best farmland in the country had already been bought up, mostly by speculators. The act simply came too late—some 77 years after Jefferson proposed it in the Congressional committee he chaired. Two-thirds of the remaining federal lands were in the dry West, where a 160-acre farm was too small to support a family. Without irrigation on a massive scale homesteaders couldn't hope to make a go of it on such a small acreage.

Despite these inherent limitations, the Homestead Act might have had a reasonable impact if the local administrators of the act had executed it effectively. Instead, they were often a subversive element working against the program and frustrating small farmers' attempts to claim land. This was especially true where administrators had been bought off by land speculators using dummy claims to acquire several "homesteads."

The government itself added a stumbling block to Jefferson's plan: it still offered land for sale at the same time that it was giving it away through the Homestead Act. Speculators with ready cash would snap up the best tracts—the most fertile land, close to markets, water, timber, and lines of communication—for low prices only to sell them later at inflated rates to settlers whose only choice was to buy the expensive land (if they had the money) or take inferior tracts—when they could get them—through the Homestead Act.

During this time the railroad industry became a major player in the land game. As an inducement to lay rail, the government turned over huge tracts of land to railroad companies parallel to where lines were laid. From 1850 to 1870 the railroads received 150 million acres—an area greater than New England—in this way.[18] The government handed over 20% of the state of Kansas to the railroad companies, and 15% of Nebraska. The railroads then sold the land either directly to settlers or to speculators who sold it in small parcels to the new farmers—after jacking up the price. The Homestead Act did nothing to slow down this trend. In fact, the railroad companies received five times as much land in the eight years following the act as they had in the preceding twelve years.

The final record of "the greatest democratic measure of all history" was less than overwhelming: Between 1862 and 1900, 372,659 land entries were perfected under the act. Just one out of every six acres distributed under the Homestead Act went directly from the government to settlers as intended.[19] The act that was to make a reality of Jefferson's vision of "a chosen country, with room enough for our descendants to the thousandth and the thousandth generation," proved most successful at enriching speculators and corrupt administrators.

The failure of the Homestead Act isn't too surprising given that many of its supporters never saw the act primarily as a means of enhancing democracy. These legislators considered the Homestead Act as a political expedient, a "safety valve" which would release the growing pressure of urban discontent in the East by providing an incentive for laborers to move west. One senator said he supported the bill "because its benign operation will postpone for centuries, if it will not forever, all serious conflict between capital and labor in the older free states, withdrawing their surplus population to create in greater abundance the means of subsistence."[20] The Homestead Act was even less successful in meeting this goal; the years following its passage were marked by increasing labor unrest in Eastern cities.

The legacy left by our early land policies is mixed. While they enriched land speculators and railroad and timber companies, they also managed to establish many farmers on the land—

certainly not as many as intended, but many nonetheless. But to say that there were many farmers misses a vital point. Throughout our history a significant, and usually underestimated, proportion of farmers have not been the independent yeomen of a Jeffersonian democracy, but tenants, sharecroppers, and debtors "owing their souls to the company store." By 1880, fully one-fourth of American farmers did not own the land they worked.

After the Civil War, a system of peonage, called the crop lien system, made a farce out of the myth of the "independent farmer" throughout the South.[21] Since cash was almost nonexistent in the former Confederate states after the war, virtually all transactions were carried out by barter. When a farmer went to buy goods from the local merchant, a running tally was kept of his purchases. For the privilege of buying on credit, farmers were charged interest often of 100 to 200%. Worse still, to guarantee payment on the purchases, farmers had to sign over their next crop to the merchant. At "settlin'-up" time, after the harvest came in, the farmer and merchant went down to the local cotton gin, where the crop was weighed and sold. The merchant received his cut, and the farmer got what was left. The problem for the farmer was that there was almost invariably nothing left after the merchant took his payment—in fact, the farmer usually still owed the merchant money.

The solution was simple: the merchant would "carry over" the farmer until next harvest—after the farmer signed over that harvest, too. It's easy to see how the system, which affected 70% of Southern farmers, soon became "little more than slavery."[22]

As the century ended and a new one began, the level of farm tenancy throughout America grew steadily. From 28% in 1890, to 36% in 1900, to 38% in 1910.[23] Even America's Heartland wasn't immune to this trend: By 1935, half of Iowa's farmers, the quintessential yeomen of Jefferson's dream, were tenants.[24] In the following year, a committee appointed by President Franklin Roosevelt declared farm tenancy the basic cause of rural poverty.[25]

Under the Roosevelt Administration, a number of programs were developed to increase chronically low farm income, ensure a stable and cheap food supply, and conserve farmland, an era which is remembered today as either the golden age of farm policy or as agriculture's dark years, depending on the political bias of the individual. Certainly the various New Deal farm programs showed a new federal willingness to regulate American industry.

Probably the most radical program of the New Deal was the land reform effort undertaken by the Resettlement Administration (RA) and its successor, the Farm Security Administration. Under the auspices of these agencies the government purchased nearly 2 million acres of land which it then rented (under favorable terms with an option to buy) to landless poor individuals. Loans were also available to small farm owners so that they could expand their operations enough to become profitable.

Even more politically ambitious was the RA's emphasis on community development as opposed to solely individual development. The federal government purchased or leased what had been Southern plantations and created whole new communities on them. Although these land reform programs faced continual court challenges and were finally ended in 1937, they were very successful. Of the planned communities, 90% were still in existence in the 1970s.[25] The measures formed the basis for a similar program of land reform which was put into effect by Douglas MacArthur in Japan after World War II.

New Deal farm programs introduced a variety of price-support measures (including government loans and crop subsidies) and production-control requirements in an attempt to raise farm income while at the same time reducing the acreage devoted to soil-depleting crops. These measures form the basis of federal agriculture programs to this day.

While the New Deal greatly advanced the cause of the family-sized farmer, its policies, like the Homestead Act of the last century, had some serious, even tragic, failings. The unintended negative effects of the New Deal were felt most harshly in the South, where the sudden removal of more than 10 million acres of cotton in a program designed to reduce overproduction led to the ruin of thousands of poor tenant farmers and sharecroppers, mostly African-Americans. Although the original legislation provided for landowners to share with their tenants the money the government paid them for removing land from production, that provision was quickly removed. In just seven years, 30% of Southern sharecroppers and 12% of tenants were forced off their land, some to be hired back as wage laborers.[27]

Throughout the country, federal supports and loans went disproportionately to the largest farmers, those with sufficient clout to cut through red tape and reap the lion's share, while smaller farmers often remained neglected. These wealthy farmers used their government checks to expand further, purchasing new technologies (particularly tractors) and buying up the farms of less fortunate neighbors.

The increasingly powerful food processing industry was able to scuttle another vital provision in the New Deal's farm program, a provision which levied a tax on food processors to help pay for the new measures. The Administration deemed it only fair that food processors be made to share some of the cost involved in bailing out farmers, since these corporations were making huge profits from the same low commodity prices that were devastating farmers. The food processors did not share that perspective, however, arguing that the 1933 tax was an unconstitutional penalty on their industry. In 1938, the Supreme Court sided with the food processors, rescinding the law and transferring the costs of the federal farm programs to taxpayers.[28]

The intrusive New Deal farm programs were anathema to succeeding Administrations of "free market" Republicans, who did their best to dismantle the agricultural bureaucracy put into place by Roosevelt. In 1953 President Eisenhower's Secretary of Agriculture Ezra Taft Benson began his tenure by offering the obligatory rhetorical homage to the yeoman farmer. "Rural people," he declared, "are a bulwark against all that aims at weakening and destroying our American way of life."[29] However, almost immediately Benson began to demolish New Deal programs designed to prevent chronic overproduction and replace them with policies that promoted increased crop production.

Benson managed this redirection of government energies in the same way that many others were advancing their causes in an era marked by Joe McCarthy's witch-hunts: by red-baiting the opposition. The Secretary of Agriculture, who later became a vocal spokesman for the John Birch Society, insinuated that the New Deal farm programs were really Bolshevism's first foray into our country. Only unlimited production could save the nation from falling beneath Stalin's boot.[30]

The policy wrapped in the flag played well in Dubuque and elsewhere in large part because it appealed to America's most cherished notions about itself. The unlimited-production route promised to "get government out of agriculture," always a big crowd pleaser. The myth of the independent yeoman, that workhorse of American political symbology, was also put to work. Ironically, it was harnessed to a farm policy that denied the need of an active federal role in the regulation and protection of farm life and economy—a policy which, if fully implemented, would destroy the yeoman farmer. For all Benson's dislike of the New Deal farm programs—which he blasted as "a sordid mess"[31]—and despite his success in doing away with production controls, United States agriculture policy when he left office in 1961 was fundamentally the same as when he began his job.

In outward appearances, that policy is little changed even today. But over the years, programs originally designed to protect the small and medium-sized farmer have instead become little more than mechanisms to funnel taxpayers' money into the pockets of the largest growers. The result has been aptly termed a system which "provided for welfare for the rich and the market system for the poor."[32]

That welfare system takes many forms, from international trade agreements which favor agribusinesses to crop subsidies, tax policies, and government loan programs. A recent General Accounting Office study found that 30% of farm subsidies go to the largest 1% of producers (the so-called "superfarms" with annual sales over half a million dollars), while the 80% of the nation's farms with sales under $100,000 a year receive less than one-third of the government payments.[33]

But the biggest beneficiaries of modern agricultural policies that push for maximum production are not the large growers, they are the businesses on either side of the farmer—those that sell farmers the chemicals and other supplies they need to boost yields (inputs), and those that buy the cheap grain assured by massive harvests (outputs). Quite often, the same company controls both inputs and outputs.

ConAgra, one of the largest agribusinesses in the United States, is a good example of this phenomenon. On the cover of its 1985 annual report is the motto "Results across the Food Chain," a slogan which sums up the corporate strategy of the food industry as a whole since World War II. In economic terms, the concept is called vertical integration, and it means that the company's goal is to control food from farm to table. ConAgra is fast reaching that goal.

On the input end, ConAgra is the nation's leading distributor of pesticides. It is also the county's number one flour miller and number one producer and marketer of frozen prepared foods, and in 1988 it became the nation's top beef slaughterer. Because companies like ConAgra are in the fortunate position of selling inputs to farmers retail and buying outputs from them wholesale, they are able to benefit doubly from policies that encourage all-out production. As farmers were going bankrupt in the early 1980s, ConAgra reported record sales and earnings each year.[34]

"These companies are giants," complained House Agriculture Committee member James Weaver. "They control not only the buying and the selling of grain but the shipment of it, the storage of it and everything else. It's obscene. I have railed against them again and again. I think food is the most—hell, whoever controls the food supply has really got the people by the scrotum. And yet we allow six corporations to do this in secret. It's mind boggling!"[35]

When an executive from Pillsbury, another of these industry giants, went before a Senate subcommittee in 1987 to criticize a plan to raise farm income by reintroducing production controls, he claimed that farmers were better served by the free market.

An angry Tom Harkin, a senator from Iowa and the cosponsor of the proposed plan, held up a piece of paper. "If you look at your return on equity last year, it was 17%," Harkin told the executive. "If farmers in Iowa could just get 2% return on equity, they'd be darn happy. They've had a negative return for the last several years."[36]

The farm crisis has been good for the food processing industry as a whole. From January 1986 to October of the same year, while farm prices fell 9%, the food processing companies posted an increase in profits of 13%.[37] In fact, these agribusinesses did better than did U.S. industry as a whole during the farm crisis. During the worst five years of the crisis, from 1981 to 1986, the fifty food processing firms listed in a *Forbes* magazine industry survey had an average return on equity of 15.1%, while U.S. industry on the whole had an average return of 12.6% during that same period.[38] But the fact that these giant corporations have been doing well while farmers have not receives little attention. When the phenomenon is mentioned at all, it is usually as an afterthought, as in this newspaper-article lead: "Cargill, Inc., the largest agribusiness in the country, has reported its highest pretax profit in 12 years despite a prolonged slump in the agricultural sector."[39]

The prosperity of Cargill et al. amidst the worst farm crisis since the Depression is seen simply as an inexplicable paradox—one of those quirks of economic life to be recognized (begrudgingly), stated (infrequently), and abandoned (quickly). But agribusiness's boom and farmers' bust dovetail neatly in a federal policy that encourages all-out production.

One reason that government policies consistently benefit these industries over family-sized farmers is that our government and these industries are intertwined. The system is much the same as it was in the late 1700s, when politicians who passed laws making land speculation eminently profitable were themselves land speculators or would benefit directly and indirectly from the passage of such laws.[40] On page 3 of the 1985 ConAgra report mentioned above is a brief note from the company's chief executive officer wishing a warm good-bye to one Clayton Yeutter. Yeutter had recently resigned from ConAgra's board of directors to serve the Reagan Administration as United States Special Trade Representative. One of his primary accomplishments in that position was to expand grain exports by keeping the price of United States grain low enough to capture more markets—a policy beneficial to Con-Agra's bottom line. In 1989, ex-ConAgra director Yeutter was named Secretary of Agriculture by President George Bush.

Probably the most egregious example of the revolving-door syndrome in the Agriculture Department occurred when Richard Nixon's first Secretary of Agriculture, Clifford Hardin, resigned under a cloud of scandal in 1971 (for allegedly increasing the price-support level for milk in exchange for thousands of dollars in campaign contributions from the dairy industry) and became a vice-president of Ralston Purina, an agribusiness giant. Replacing Hardin as Nixon's Secretary of Agriculture was Earl Butz— recently a director of Ralston Purina.

Butz, a protégé of Ezra Benson, more than any other single person was responsible for the boom era of the 1970s that led to the crisis of the 1980s. Like Clifford Hardin, Butz was no stranger to scandal. He was eventually forced to resign on the eve of the 1976 presidential election after making a vulgar racist joke. In 1981 Butz served a 25-day jail term for tax evasion. He managed, however, to evade responsibility for what was probably the most serious scandal of his career: the Soviet grain deal of 1972.[41]

Due to the combined effects of a rising world population, an increased demand in developing nations for meat (coming from cattle raised on grains), and two years of bad weather, the world was experiencing a critical shortage of grain in 1972. It was against that backdrop that the United States sold the Soviets a record 19 million metric tons of grain—the largest single agricultural transaction in world history—at bargain-basement prices, thanks to three-quarters of a billion dollars in low-interest loans provided by Butz's Department of Agriculture.

While the grain sale proved to be a windfall for agribusinesses, it benefited grain farmers to a lesser extent, and it added a billion dollars to the nation's food bill.[42] Due to what a Senate Agriculture Committee staff memo later called "coziness between the Department of Agriculture and private grain exporters," these giant companies received $300 million in unnecessary government subsidies in the deal.[43]

A Justice Department investigation failed to prove Butz guilty of any conflict of interest in the deal, but the odor of corruption stayed with the Secretary until he resigned—an odor that wasn't lessened when Assistant Secretary of Agriculture for International Affairs Clarence Palmby, the man who led the U.S. negotiations with the Soviets on the grain deal, resigned his $40,000-a-year government post soon after the agreement was concluded to take a $60,000-a-year position with Continental Grain Company—the corporation that had made the single largest sale of grain in the deal.[44]

But the revolving door is just one of many weapons in the armory of any giant corporation. The political action committee (PAC) contribution is another. In one recent election cycle, food-industry PACs wanting to keep farm chemical usage up outspent environmentalist PACs 18 to 1—contributing $6.5 million to Congressional candidates ($1 million of which went to members of agriculture committees), compared to the environmentalists' paltry $362,000 in donations.[45]

High-paid agribusiness lobbyists (many of whom have been through the revolving door themselves) effectively press the industry's many cases behind the scenes, wining and dining legislators and bureaucrats when bills or regulations affecting their interests are considered. There were so many lobbyists from the farm chemical industry "working" the Environmental Protection Agency's headquarters in Crystal City, Virginia, at one point that EPA staffers dubbed them "hall crawlers."

Of course there is more to the rise of agribusiness and the fall of the family farm system than simple corruption and the venality of individuals. Changing markets, technological advances, economies of scale—all of these considerations are both real and important. But the influence of giant agribusiness is felt even in these supposedly neutral areas as well. For example, conventional wisdom has it that improved technologies coming out of land grant institutions inevitably lead to larger and larger farms and fewer and fewer farmers. Yet research at agricultural colleges is just as subject to influence as is the legislative process.

"There is nothing value free about research that consistently benefits the fewest and the biggest," says attorney William Hoerger, who represented a group of family farmers in a 1973 lawsuit brought against the University of California, the country's premier land grant system. The group California Action Network (CAN) charged that the university was misusing federal funds—which CAN maintained were intended to benefit small farmers, consumers, and rural residents—by conducting research that benefited primarily agribusinesses.[46] As an example, CAN cited the case of the mechanized tomato harvester designed at the University of California in the 1960s. The average size of a California tomato farm was 35 acres in 1964, when the first mechanical harvester was developed, but the machine was designed to be used with a minimum of 75 acres and to reach peak efficiency at 150 acres. Eight years after the harvester's introduction, the average size of a California tomato farm was 363 acres, and the number of these farms had dropped from 4,000 to 600. The machine, critics calculate, caused the loss of 18,000 jobs for fieldworkers.[47]

"Technology per se isn't the problem," says Walter Goldschmidt, professor emeritus of anthropology and psychiatry at the University of California and author of *As You Sow*, a groundbreaking 1947 study of the social and economic effects of agribusiness on two California rural communities. Goldschmidt tells of a visit to a Japanese farming village in which he found diesel-powered mini-plows, mini-planters, and mini-harvesters in use on one-acre rice paddies.

"Whether a piece of machinery is designed to benefit the small farmer with only one acre, as in Japan, or the large grower with tens of thousands of acres, as in California, is up to the designer," says Goldschmidt.

And what prompts scientists to pursue research that will benefit agribusiness while ignoring projects that will help smaller farmers? Very often, that decision turns on private industry funding. Although private grants account for only a small percentage of the total funds involved in research at land grant schools (only between 5 and 10%), these grants are critically important to researchers. This is because while public money pays for virtually all overhead expenses—including laboratories, salaries, etc.—researchers depend on private grants for "cash costs" such as seeds, chemicals, machinery, and, perhaps most importantly, to pay graduate students to help on projects. Private monies make up the bulk of "flexible funds" used by researchers to pursue specific projects.[48]

Given that agribusinesses have the funds to commission research and small farmers do not, it is no wonder that technological advances most often benefit the former to the detriment of the latter.

Another loser under post-World War II farm policies has been the consumer, despite the fact that proponents of the policies claim that the consumer has been the main beneficiary of these programs because of the availability of cheap food. "Americans spend a smaller portion of their incomes on food than do citizens of any other country," these advocates claim. But our food only appears cheap because the price we pay at the grocery store is just a fraction of the total cost. In reality, through a variety of hidden costs we pay dearly for our "cheap food."

First, American taxpayers foot the bill for federal farm programs, paying billions every year for farm subsidies, crop deficiency payments, drought disaster aid, and federally funded agricultural research programs. None of these costs is figured in when agribusiness advocates boast about the nation's inexpensive food supply.

The public also pays for this "cheap food" in the diminished quality of the environment that results from modern agricultural practices. And then there is the multi-billion-dollar cost of bailing out lending institutions, both federal and private, sinking under the weight of nonperforming agricultural loans. And the money spent in job retraining programs for farmers pushed off the land, and the additional money for food stamps and for other welfare programs to help ex-farmers who can't find new jobs and so exist on the fringes of society.

But the biggest immediate losers from agricultural policies that push maximum production at any cost are the very individuals these programs are supposed to help: middle-sized family farmers like the Bolins, and the rural communities they inhabit. From 1981 to 1987, 26,000 Iowa farmers—about 20% of the total—went out of business.[49] But the family farm—never the robust institution of popular mythology—has been in serious, and many say fatal, trouble for decades. Our farm population has plummeted from 30 million down to 5 million since the 1940s, while the average farm size has more than doubled during the same period.

The change in agriculture is even more dramatic when looked at in terms of sales. A large percentage of the remaining farms produce a small amount of food and are often referred to as "hobby farms." Today, just under half of the nation's food is grown on the largest 4% of farms; one-third is produced by the largest 1%. A recent government study anticipates that just 50,000 of the largest farms will account for three-fourths of all agricultural production in this country by the year 2000.[50]

According to many farm activists, like the Reverend David Ostendorf, director of the Iowa-based farm advocacy group Prairiefire, the most significant and disturbing trend in agriculture has been in the concentration of land ownership. "A widely dispersed base of land ownership is fundamental to American democracy," says Ostendorf. "Look at what's happened in Central America because of this kind of concentration. If things continue on as they are now, I'm convinced we'll become a little Central America, U.S.A., here in Iowa."

There is little concern in this country over the fact that the top 5% of American landowners own 75% of our land, and the bottom 78% own just 3%.[51] The figures are even more unbalanced in many regions. For example, the top 5% of landowners hold 90% of privately owned land in Hawaii, Florida, Wyoming, Oregon, and New Mexico, and hold 80% of it in Washington, Utah, New York, Nevada, Maine, Louisiana, Idaho, California, and Colorado.[52] In looking at these figures, it is instructive to note that on the eve of the revolution in Cuba, the largest 9% of all landowners held 62% of that country's land, while 66% owned only 7%.[53]

The insurance industry is a little-known but major player in the land ownership game in this country. In 1986, insurance companies owned $2.4 billion worth of farmland, double the amount held the previous year. One company, Prudential, owned 800,000 acres in 18 states.[54]

"During the Great Depression, insurance companies accounted for two-thirds of the farm foreclosures, and they ended up with 15% of America's farmland," says Ron Kroese, director of the private research group Land Stewardship Project. "Only 2% of that foreclosed property was returned to the original owners. That was also the watershed period where we saw the number of farms decrease and the average farm size increase dramatically. That's our concern today."

Activists like Kroese and Ostendorf have reason to be concerned. Only 58% of American farmland sold in 1986 was bought by farmers. Farm management companies, which often hire bankrupt farmers to work the land they once owned, for hourly wages, increased their control over agricultural land by 36% between 1980 and 1986.[55]

Concentration within the food processing industry is even more extreme than in the field of land ownership—and at least as harmful to family farmers. A mere 0.25% of food processing companies today control two-thirds of the industry's assets.[56] These corporations use their economic muscle to keep the price of raw materials (unprocessed food) low. There has been a similar concentration in the farm input sector—those industries that sell farmers the chemicals and hardware they need to grow crops.

Caught between these giants selling farm inputs and the giants buying farm products (often—like ConAgra—the same company), the farmer's share of the consumer's food dollar has plummeted to its lowest point since the government began keeping figures in 1913. In the early 1970s, a loaf of bread sold for 28 cents. Of that price, a farmer received 4 cents. Today, the price of a loaf of bread has more than doubled, but the farmer still receives less than a nickel.[57]

As a group, African-American farmers have suffered disproportionately during the farm crisis, losing their land at a rate ten times faster than their white counterparts.[58] In fact, black farmers have been in dire straits for years; but when they were being squeezed off their land, it was of interest to only a few scholars. "Trends" happen to blacks, "crises" only to whites. "The farm crisis began for black farmers when blacks began farming," says Jerry Pennick, director of the Land Assistance Fund of the Federation of Southern Cooperatives in Atlanta, Georgia. "It's just gotten worse during the general farm crisis."

In 1920, there were 926,000 farms operated by African-

Americans in the United States, most in the Southern states of Mississippi, the Carolinas, and Texas.[59] One of every seven farmers at that time was black. By 1982, the total number of African-American farmers had dropped to 33,000; only one farmer in sixty-seven was black. Even in the South, in the so-called "Black Belt" where black farmers were at their strongest, the weight of the farm crisis has fallen harder on African-Americans. In 1964, there were 58 Southern counties in which black farmers were the majority. Today, white farmers outnumber blacks in every county in the United States.

"Today, blacks are losing their land at the rate of 9,000 acres each week," says Gary Grant, a community organizer in Tillery, North Carolina. "If this trend continues, blacks will soon be a landless people. And a landless people is a powerless people."

In addition to all the problems faced by white farmers, African-American farmers have to contend with the additional burden of racism. That racism comes not only from local bankers and business leaders but also, according to black farm activists like Shirley Sherrod, from the government itself.

"The government is supposed to be a force for keeping blacks on the land, but it hasn't been at all," charges Sherrod, a field coordinator with the Federation of Southern Cooperatives. "It has worked against us much of the time."

In fact, the Department of Agriculture has a long history of racism. African-American farmers have always had a hard time obtaining loans from USDA lending agencies. In 1984 and 1985, the USDA lent $1.3 billion to farmers nationwide to buy land. Of the almost 16,000 farmers who received those funds, only 209 were African-Americans. The USDA's Extension Service has for decades discriminated against blacks by underfunding programs to black colleges and black 4-H groups.

"The Extension Service hasn't taken the initiative to transfer new technology to the black farmer, and so he's remained behind," says the Land Assistance Fund's Pennick. "They've never given anything but token help to black farmers. And not just under the Reagan Administration. Every Administration has put the black farm crisis on the back burner."

Within the USDA itself, which was strictly segregated until 1965, 100 African-American employees recently charged that they continue to face discrimination, citing the fact that while 118 Foreign Agricultural Service employees hold coveted overseas positions, there are only 4 African-Americans in their ranks.[60] The USDA has not usually dealt kindly with black employees who have rocked the boat. In 1986, Dr. Edith P. Thomas, a nutritionist in the USDA's Extension Service (who also happened to be the highest-ranking black woman in the department), charged that the USDA was denying nutrition assistance to poor African-American families. One week later the USDA moved to fire her.[61]

The little-noted disappearance of the black farmer has serious implications for blacks both rural and urban. As poor rural blacks have been pushed off the land, they have tended to migrate to urban centers in the South, creating new ghettos there.[62] The trend is strikingly similar to the migration of poor rural African-Americans who were forced off their farms in earlier decades, and who moved into urban ghettos in the North. Only time will tell if these new Southern ghettos will explode as their Northern counterparts did in the 1960s. We have apparently failed to grasp the significance of one presidential report's conclusion that "the urban riots during 1967 had their roots, in considerable part, in rural poverty."[63]

Rural Southern communities are themselves hurt, both economically and politically, by the loss of a widespread class of African-American farm owners. These landowning blacks are more likely to vote than are nonlandowning African-Americans in the same community. Black farm owners acted behind the scenes during the civil rights movement, putting their farms—which for most represented a lifetime of work and saving—up as bail for activists who went to jail during those years. Southern black farmers, the generally uneducated and always poor tillers of the hardscrabble soil, were among the unsung heroes of the civil rights movement.[64]

Today, those same farmers are leaving the land, both through foreclosure and simply because a majority are too old to continue farming. In 1982, 58% of African-American farmers were above the age of 55, with a majority of that group over 65 years old. By comparison, only 41% of white farmers were over 55, with the majority of that group clustered around the younger end of the spectrum.[65]

The figures are even more racially unbalanced when the number of young farmers in the country is considered. The percentage of white farmers under the age of 35 is more than double the black rate: 15.9% for whites and 7.8% for blacks.[66]

No matter why black farmers leave farming, the result is the same: with few young blacks able to obtain the financing necessary to start farming these days—and few even wanting to enter a career in which the odds seem so stacked against them—the land ends up in white hands and the day draws closer on which the phrase "African-American landowner" will be an oxymoron.

The effect of modern agriculture on America's Heartland is particularly evident when the land itself is considered. From the lofty perspective provided by a jet cruising at 30,000 feet, the arid lands of southwestern Nebraska reveal both the best and worst capabilities of a people seduced by its own cleverness. Nowhere is the old pioneer dream of "making the desert blossom like a rose" more visibly fulfilled than it is here, directly below the cross-country traveler. Hundreds of perfect circles of the deepest emerald green, each one a half mile to a mile in diameter, are described against the dry brown, almost white chalky soil that is common throughout the western fringes of the Great Plains. The circles are actually fields of corn growing in land that was formerly suitable only for grazing cattle. The transition from rangeland to farmland was made possible by the development and spread of center-pivot irrigation—an ingeniously simple system that brings water to thirsty crops through quarter-mile-long sprinkler arms which rotate around a central point like the hands of a clock.

While the trundling arms have been unglamorously referred to as overgrown lawn sprinklers, their impact on dryland farming has been tremendous. Between 1972 and 1982, center pivots helped Nebraska boost its corn production by 44%.[67] Today, more than a million acres of once-dry Nebraska farmland are watered by center pivots, part of an irrigation system that adds billions of dollars to the state's economy each year. As more than one Cornhusker has observed, irrigation *is* the economy in this part of the state. From six miles up, the dark circles of fertile cropland stand out against the pale and miserly earth in a truly impressive display of American ingenuity triumphing over nature.

If it were somehow possible to burrow deep beneath the earth, however, to go far below those new fields framed by dry ground, a very different picture would present itself, a picture suggesting not American ingenuity, but American greed and short-sightedness. Below the fields, the source of that life-giving water is rapidly being sucked dry.

The depletion of the Ogallala aquifer—once the largest underground body of fresh water in the world—is in itself something of
a marvel, an accomplishment (if that word can be used in discuss-
ing so odious a deed) whose magnitude rivals the construction of
the Great Pyramid at Gizah. The Ogallala, which lies beneath
parts of eight states, was formed during the Pliocene and early
Pleistocene epochs by water draining eastward from the Rocky
Mountains. In its cake-like layers of sand, gravel, and porous
rock, the aquifer once held up to 3 billion acre-feet of water[68]—
more of the precious liquid than America's greatest river, the
Mississippi, carries to the Gulf of Mexico in six years. But if farmers continue to draw water from the Ogallala at the present rate,
that prehistoric reserve, which today waters one-fifth of all irrigated cropland in the country,[69] is expected to run out in as few
as 40 years.[70]

It is hard to imagine such a vast body of water being used up in
such a relatively short time. The Ogallala wasn't even put into
widespread use until the mid-1930s. But dryland farmers have a
voracious appetite for water. Land barely capable of producing 50
bushes of corn per acre can, with irrigation, be coaxed into growing 115 bushels.[71] Farmers' appetite for water is further stimu-
lated by government programs which make corn production
lucrative even on the driest marginal lands of the Great Plains. As
a result, the number of center-pivot systems in Nebraska (which
can each pump up to 1,000 gallons of water a minute) grew from
fewer than 3,000 in 1972 to around 25,000 just a decade later.[72]

Aquifers such as the Ogallala are continually being re-
plenished—"recharged," in scientific terminology—by snowmelt
in the late winter and by rain percolating down through the soil
during the rest of the year. The problem comes when water is ex-
tracted faster than it is naturally replaced. The result is called an
"overdraft"—a term that is all too familiar to anyone with a checking account. The Ogallala is running dry simply because irrigation
withdrawals result in tremendous overdrafts—14 million acre-feet
of water a year (an amount roughly equivalent to a year's flow of
the Colorado River) in the Texas-Oklahoma region alone.[73]

What is taking place on those thousands of fertile circles that
seem so impressive to the jet-borne traveler can't truthfully be
called farming. Hydrologists have a better word for the activity;
they call it groundwater "mining." The process is to the communities of the rural Midwest what strip-mining is to the people

surrounding the coalfields of the Southeast: a short-term boom and a long-term bust.

Our system of agriculture doesn't just mine water; the soil itself is rapidly disappearing. Every year in the United States we overdraw our soil account—eroding more topsoil than nature can replace—to the tune of 1.7 billion tons.[74] That is the equivalent of losing two Rhode Islands annually. Over time, such profligacy adds up. In the 200 years since our country's founding we have lost fully one-third of our total cropland topsoil.[75] In Iowa, home of some of the world's most fertile soil, it's taken a little over 100 years to use up half of the state's topsoil.[76] An Iowa farmer who doesn't practice sound conservation techniques loses over two bushels of rich black topsoil for every bushel of corn produced,[77] and it is by no means rare to find losses double that amount.[78]

Soil overdrafts of this magnitude are extremely costly—even for a nation born with the world's largest soil "account." In the short term, topsoil carried off by water and deposited in rivers and ports must be removed by expensive dredging operations. The migrating soil also increases the likelihood and severity of floods by reducing river and reservoir capacity. Off-field damages caused by cropland soil erosion cost this country an estimated $2.2 billion a year.[79] Soil erosion also reduces crop productivity. A corn field which has lost one inch of topsoil, for example, will produce 6% smaller harvests.[80] Reduced productivity due to soil erosion costs farmers (and consumers) $1.3 billion annually.[81]

The long-term consequences of soil mining are even more serious, of course, because for future generations the problem will not be one of increased cost, but of survival.

Agriculture is also mining another natural resource to the limit: petroleum. Proponents of modern agriculture often brag that one farmer in the United States feeds 116 people, but that claim is something less than the truth. It would be more accurate to say that one farmer, *backed up by huge amounts of fossil fuel,* feeds 116 people, for farming is the nation's largest consumer of petroleum products. It is dependent on oil for an amazing variety and quantity of farm inputs, from pesticides to fertilizers to diesel fuel used by tractors, combines, and irrigation pumps.

The replacement of labor by oil (and the capital necessary to purchase the stuff) is one of the keys to modern farming. All across the country, millions of Farmer Browns have been replaced by millions of barrels of crude. The problem facing future gen-

erations is that while Farmer Brown is a renewable resource, oil is
not, and a system of agriculture dependent on the latter is not
sustainable in the long run.

Just as important as what agriculture takes out of the earth is
what it puts back into the environment, and farmers today are
putting an amazing amount of chemicals—many of them highly
toxic—into the nation's rural waterways, soil, and air. In some
cases, these dangerous chemicals are finding their way into our
drinking water and onto supermarket shelves.

In 1955, farmers used 2 million tons of nitrogen fertilizers;
today they use 12 million tons.[82] The growth in the use of pesticides has been equally dramatic, nearly tripling between 1965 and
1985.[83] Today, U.S. farmers apply 390,000 tons of pesticides a
year—or around three pounds for every man, woman, and child
in the country.[84] There is no question that these chemicals have
helped farmers to produce more crops per acre under cultivation.
But at the same time these chemicals pose a threat to human
health, as well as to the health of other animals, as biologist and
nature writer Rachel Carson pointed out in her 1962 book *Silent
Spring*. Although health problems caused by farm chemicals, particularly by pesticides, were known within the scientific community before *Silent Spring* appeared, it was Carson's work that
brought that information out of the laboratories and into public
view. The outcry sparked by her work led to the banning of one
of the most widespread and damaging pesticides (DDT), the formation of the Environmental Protection Agency (EPA), and the
birth of the U.S. environmental movement.

It is generally forgotten today, but her book, which is now re-
vered as a "classic," was widely vilified when it first appeared. *Time*
magazine called *Silent Spring* an "emotional and inaccurate outburst." A chemical-industry magazine branded the book "science
fiction," likening it to "The Twilight Zone." Ad hominem attacks
fell like acid rain: "I thought she was a spinster," commented a
member of the Federal Pest Control Board. "What's she so worried about genetics for?"[85]

But Carson's facts couldn't be so easily dismissed. Unfortunately, neither could the influence of the farm chemicals industry.
Many of the same compounds Carson criticized more than a quarter century ago are still in widespread use in the United States and
abroad, and many new formulations have been added to agriculture's chemical arsenal.

Because of their proximity to food-raising areas, rural people run a greater risk of developing cancer from exposure to pesticides than do consumers in general. The problem of groundwater contamination by farm chemicals is particularly serious in rural America, where 97% of the population depends on underground aquifers for its water supply.[86]

Recent studies of groundwater supplies in rural areas have yielded sobering results: 17 different pesticides were detected in the groundwater of 23 states in a 1986 EPA study.[87] In Iowa, pesticides have shown up in one-third of the state's wells; in Nebraska, one-half of the state's municipal water systems are similarly tainted.[88] A California study found 58 pesticides in 3,000 wells spread across a 28-county area. Half of those wells had to be closed.[89] The problem is all the more serious since groundwater pollution is almost impossible to clean up due to the fact that aquifers are virtually inaccessible.

Despite a lack of data directly linking pesticides with cancer in humans, the case against many of these chemicals is on a par with the one against cigarettes: there may not be a smoking gun, but there are a lot of bodies lying around, and the smell of gunpowder hangs heavy in the air.

Several studies have found correlations between elevated cancer rates in rural areas and the use of farm chemicals. One 1983 study determined that residents of Iowa counties in which large quantities of pesticides were used were 60% more likely to die of leukemia.[90] Other studies have found that farmers who use pesticides are more apt to develop Hodgkin's disease, non-Hodgkin's lymphoma, and cancer of the skin, lip, brain, stomach, and prostate.[91]

Perhaps the single most startling development of agriculture's chemical era is the recent appearance of "cancer clusters," rural communities in which cancer rates soar far above normal.[92] McFarland, a farm community of 6,400 located in California's fertile San Joaquin Valley, is one such "cancer cluster." Between 1981 and 1984 public health officials found 11 children with nine different kinds of cancer in McFarland—about four times the expected number of cases. Eighty miles to the north, in tiny Fowler, a community half the size of McFarland, officials discovered that 3 children developed leukemia during the same time period. Residents of both towns also reported a higher than normal number of miscarriages, low-birth-weight babies, and fetal deaths.

Although these health problems have not been definitively linked to the heavy use of farm chemicals in the area, there is good reason to suspect that such a link exists. The city water supply in McFarland is so contaminated by high nitrate levels (the result of the use of nitrogen fertilizer) that for over 20 years consumer water bills have contained a warning that the water is not fit for infants to drink. Although McFarland's aquifer tested negative for the presence of pesticides, health officials pointed out to residents that their tests were only capable of detecting a few of the 15,000 pesticides used in California. In Fowler, residues of two pesticides, both carcinogens, were found in the water supply. Significant amounts of one of these, 1,2- dibromochloropropane (DBCP), had been detected in area aquifers in 1979 and had forced the closing of more than 100 wells. Three years later, a county study found above-normal cancer rates in communities that had high DBCP levels.

"What we see in McFarland and Fowler may be only the tip of the iceberg," concluded California Lieutenant Governor Leo McCarthy.

Of all rural residents, farm workers and their families are most at risk. They labor all day in fields sprayed with toxic chemicals, the water they drink is more likely to be contaminated with them, and pesticide residues cling to their clothes when they come home at night. "They're really living in a pesticide-saturated environment," says Dr. Marion Moses, assistant clinical professor at the University of California School of Medicine in San Francisco and an expert in environmental and occupational medicine. Moses has been studying the effect of pesticides on California fieldworkers and their children since the mid-1960s, and travels around the country speaking about what she calls "the big dirty secret of agriculture": the widespread use of child labor in pesticide-laden fields.

"Twenty-five percent of the work force is under the age of 16," she maintains. "Anywhere the wages are low, children are in the fields—I see it all the time. They have to work in the fields to bring money into the family. We know more about farm machinery than we know about the health effects of pesticides on kids."

At least 3 million people work as farm laborers each year, one-sixth of them on so-called "superfarms" in California, Florida, and Texas.[93] (Other estimates put the total number of fieldworkers at closer to 5 million.)[94] Because day care is virtually nonexistent for fieldworkers, even children too young to work (i.e., children under the age of six or seven) must accompany their parents into the fields, where they spend the day playing and sleeping surrounded by hazardous chemicals. With no other water around, farm workers and their children are often forced to wash in and drink out of heavily polluted irrigation canals.[95] Farm workers are not protected by OSHA regulations; they must depend, instead, on the good graces of the EPA itself for protection. The EPA, however, has two sets of safety standards: one for the public at large (a standard which is itself highly questionable) and a more relaxed one for farm workers. The agency generally restricts or bans the use of chemicals which cause one new case of cancer for every million citizens exposed. But it allows ten times as many farm workers to develop cancer before drawing the line—one in 100,000.[96]

This profligate spending of natural resources—at the expense of future generations—could perhaps be justified if it were necessary to our survival. But necessity has not prompted our actions. Our soil is washed away and our rural water supplies are permanently polluted for the basest of reasons: the pursuit of the quick buck. Even as corn gluts the market, government policies encourage farmers to pour on pesticides and fertilizers to boost production. These same policies push farmers to expand the number of acres devoted to corn, often onto highly erodible land—and in the case of Nebraska, Texas, and Kansas, onto irrigated land. The result is the wholesale squandering of soil, water, and human health throughout the rural Heartland.

The short-term financial crunch of the 1980s called the farm crisis, though diminished for many farmers, is by no means over. A significant percentage of farmers carry dangerously high debt loads. If crop subsidies are cut back, or simply dropped altogether—as many are demanding as a way to fight the federal budget deficit—another wave of foreclosures is almost certain to sweep across rural America.

But even short of such a catastrophe, the long-term problems caused by an agricultural policy that pushes farmers to produce more and more crops in a world already awash in grain will inevitably destroy what is left of this country's family farm system. Attrition, not foreclosure, will in that case be the force that does the family farm in. The destruction will happen more slowly per-

haps, but just as certainly. And it will occur with less public scru-
tiny, minus the telegenic drama of farm families huddled together
tearfully at forced auctions.

Such a scenario is, in fact, about to be played out, warns sociologist Charles Geisler, who has coined the word "landshed" to
convey the magnitude of the imminent transfer of some 300 million acres of farmland now in the hands of older farmers—
farmers who will soon retire.

"It's difficult to say who will end up with that land," says Geisler.
"It's almost certainly the last round of transfer in which the land
is going to go from farmer to farmer using any kind of liberal
definition of a family farm. But one thing is certain: that land is
going to move. And we've only begun to think about the implications of that transfer for the larger society."

What happens to the people who formerly owned and farmed
the land? To the communities that depend on the trade of these
farmers? To the Heartland itself, which, however imperfectly,
came the closest of any region to fulfilling Jefferson's dream of a
nation of independent yeomen? What, in short, happens when
the farm crisis comes to town?

3

The Rise of
the Rural Ghetto

*The social philosophy of private accumulation is a lot like
Calvinism: an elect few are chosen to live in paradise,
while the rest can go to hell.*

DONALD WORSTER
in *Meeting the Expectations of the Land*

Tipton, Iowa

Grace Countryman first suspected something was up the day
Verne Abbott, the store's manager, got the phone call from
headquarters.

"Well, the girls figured maybe that was it," says Countryman,
a short, pleasant woman in her sixties who has worked the cash
register at Schultz Brothers Variety Store for over 14 years.
"And then a salesman came in and said he had to pick up some
stuff because he got the word that we were going out of business. Then we really knew it was over."

Countryman punctuates each sentence with a small laugh
that sounds more like a cough. She laughs when she tells how
she found her job here back in 1975 through her daughter,
who worked part-time at the store after school; when she recalls
how friends would drop in on rainy afternoons with the excuse
of buying a dish towel or a pair of nylons; about the irony that
places the store's closing on April Fools' Day. She even laughs
as tears fill her eyes and she talks about the future.

"Oh, I'm sure I'll feel real let down when it closes," she says. "I hope I can find something else where I can be around people. I like that. I hope I can find anything, really. My husband retired just last month—we didn't know this was coming. So, you see, we were sort of counting on my wages here. I'm older than some of the girls and I know it'll be a little more difficult for me to find something." She looks down at the floor and laughs.

Schultz Brothers Variety Store has occupied the first floor of the yellow brick Cobb Block Building just north of the grass lawn and wooden benches of Tipton's town square for 58 years. Racks of clothing stand to the left of the glass-door entrance; school supplies, household goods, toys, and hardware line shelves and displays to the right. The familiar smell of Spic and Span is everywhere.

But the signs of change are unmistakable throughout the store. Some literal: hand-lettered notices declaring "No Lay-Aways, No Returns, No Checks" are posted every few feet. Large yellow and green letters hanging slightly askew in the front window declare with an unemotional simplicity typical of Iowa and its people, "Good-bye Tipton."

Some signs are not spelled out in words, but they are just as plain. Broken-down shelving lies in stacks to the rear of the store, separated from what's left of the merchandise by a rope like those used by the police to cordon off accidents. With the growing heaps of metal and glass strewn over the floor, the store looks like an accident in progress. Each day, the rope that separates the store's full past from its empty future moves a few feet forward.

Verne Abbott, Schultz Brothers' 31-year-old manager, is in charge of the closing-up operation. He hustles down the aisles, like a ball boy at center court, Wimbledon, rearranging the dwindling stock, moving signs, slashing prices. Abbott is the only one of the store's employees who will stay with the company. After April 1, he will become comanager of the Iowa Falls store.

"Frankly, I was shocked when I got the phone call," says Abbott, who is small and delicately boned, with heavy dark-rimmed glasses that make him look very much the store manager he has been for four years—two of them in Tipton, another two at the Schultz Brothers store in Anagal, Wisconsin ("home of the Metz fishing lure," he adds with a high-pitched, gently mocking laugh).

He is seated in a cluttered crow's-nest office, a loft tucked into a back corner of the store. "That's probably the best way to put it: I was shocked. The employees were shocked when I told them. When the rest of the town found out, they were shocked, too."

Not that it hadn't been a long time in coming. Sales had been dropping for several years as farm income declined. Other stores around the square had folded up like mobile homes before a tornado: the bridal shop, a couple of restaurants, a jewelry store. Then, two years ago, came the coup de grace: the giant discount chain Wal-Mart bought some land just north of town. Within weeks a 50,000-square-foot store with a parking lot ten times bigger than the entire Schultz Brothers store obscured the former corn field.

"Well, that didn't help us any," says Abbott with a laugh as bleak as Countryman's. "No store can compete with them the way we tried. We cut our prices to the bone. And I mean to . . . the . . . bone." Abbott taps out the words with his forefinger on the desk top before him as though in Morse code. He leans back in his chair and sighs, running his hand through straight dark hair.

"But they have over 1,000 stores. They advertise on TV all the time. There was just nothing we could do."

To be precise, Wal-Mart has 1,345 stores in 25 states, 19 of them in Iowa. It is the country's third largest retail chain (behind Sears and K-Mart), with sales totaling $20 billion in 1988. The fastest-growing company in its field, Wal-Mart is expected to challenge top-dog Sears by 1993.[1]

Retailers across rural America share Abbott's feeling that there is nothing you can do when "Wally-World" (as some of its detractors call it) moves to town.

"Competing stores try to hang on three, four, or five years," says economist Kenneth Stone, who has studied the impact on surrounding businesses when Wal-Mart opens a new store. "They find they can't do it any more, and they close their doors." Even distant merchants are not immune to the changes wrought by the retail giant. Stone found that a Wal-Mart store lures away as much as $200,000 annually from small towns within a 20-mile radius. Only retailers willing and able to exist in the cracks left in between Wal-Mart's 30-odd departments may survive.

"When Wal-Mart opens a store in a town, the small business is going to have to adjust," a Wal-Mart spokesperson said recently. "If they're willing to adjust, their business will be stronger."[2]

Stone's study bears out Wal-Mart's claim: some businesses can actually benefit from the giant's presence—if they're savvy enough to restructure their store so that it doesn't compete directly with the new store. But for many businesses such a transformation is impossible. You can't easily go from running a hardware store—which competes with Wal-Mart—to operating a restaurant—which doesn't. And so the process termed "the Wal-Marting of rural America" continues.

Abbott and his staff tried to compete in a variety of ways with the giant out on the edge of town. They instituted a children's ID program in cooperation with the county sheriff's department as a public relations gesture, and supplied candy and the bunny suit for the annual Easter egg hunt at the local senior citizen's home. Wal-Mart dashed off a $1,000 check to help the home buy a new whirlpool. Abbott ran sales constantly, special-ordered anything a customer wanted, started issuing weekly circulars in the local paper, and refunded money cheerfully and without question if someone decided there weren't enough holes in his new salt shaker or that she didn't look as good in red at home as she had at the store. But how do you compete with a store that offers 100 different kinds of shampoos and conditioners, all selling for less than the eight or ten brands you carry, and that does the very same thing with dishes, motor oil, diapers, ratchet wrenches, batteries, sunglasses, paper clips, romance novels, licorice, aspirin, wrist watches, charcoal briquettes, boots, T-shirts, and everything else you stock (and a thousand things you don't) and then advertises the hell out of its "low-low-everyday" prices on TV and the radio and in the newspaper?

"You don't," says Abbott. "You can't."

Abbott's battle was doomed from the start. It was an economic replay of the conquest of the Americas in which the Indians' bows and arrows were up against the conquistadors' muskets and cannons.

As the publisher of a weekly newspaper in the nearby community of Anamosa, John Adney had a chance to watch this unequal battle up close when Wal-Mart opened its store there in 1984.

"I originally wrote editorials in favor of them coming to town," he recalls. "In fact, I was a member of the Chamber of Commerce group that went down to the Arkansas headquarters to convince them to move to Anamosa."

The new business promised jobs, Adney explains, an expanded selection of goods, and something less tangible but even more important to the community: The firm's decision to locate in Anamosa was seen by residents as an affirmation of their town's vitality and the harbinger of a bright future. If a big company like Wal-Mart had selected Anamosa as the site for a new store, the thinking went, than other businesses would soon follow.

Recalls Adney: "Wal-Mart received plenty of free publicity as the paper rightfully reported on the new business—from the selection of the building site to the ribbon-cutting, complete with photographs of smiling corporate executives and local civic leaders. Our honeymoon didn't last. When established in Anamosa, Wal-Mart began what I would consider an advertising war between my paper and the local advertising shopper. I was convinced by store managers and home-office advertising executives to grant special lower rates to Wal-Mart to undercut prices offered by the shopper."

After playing one publication against the other for a time, the giant retailer stopped advertising in Adney's paper. "I felt used by them. I was burned real good," he says. Since then, Adney has watched several downtown stores close. Some of them, he says, were simply poorly managed, and only survived before Wal-Mart's coming because people had little choice about where to shop. "Frankly," he says, "they deserved to close."

But, like the rain, the weight of the giant discount store falls equally on good and bad managers. Anamosa also lost what Adney considered a well-run and very popular department store. "They just couldn't compete with Wal-Mart."

And despite all Verne Abbott's efforts, neither could Schultz Brothers Variety Store. As Abbott looks around the near-empty store, he thinks about the changes that lie ahead for Tipton. With the demise of Schultz Brothers and the coming of Wal-Mart, elderly Tipton residents face an immediate problem: Schultz Brothers was within easy walking distance—right off the town square. The new Wal-Mart store, however, is located beyond the far north edge of town. To shop there requires a car, and many of the older people do not drive.

The closing of Schultz Brothers will alter the habits of *all* residents who were used to shopping there and then strolling around the town square, window-shopping, running into friends, maybe stopping in at the M&L cafe for lunch or coffee. For nearly 60 years that was the flow of life in Tipton. Some argue that the "one-stop shopping" Wal-Mart provides is a more efficient way of going about the business of living, and there is little doubt that, strictly speaking, they are right. What is less certain, however, is whether what Tipton residents will gain in efficiency makes up for what they will lose in community identity, lost businesses, and the demise of shopping on a human scale.

"This sure puts a hole in downtown," Abbott finally concludes. "You know, Schultz Brothers has always been a downtown anchor in Tipton. Maybe *the* downtown anchor."

The ocean simile again—just as common and just as true today as it was when pioneers came to this land of tall grasses and open skies in wagons called prairie schooners. Strolling around Tipton, down drowsing streets, passing one boarded-up business after another, it is easy to see Abbott's point about Tipton losing its anchor. And it is easy to imagine the town as if drifting, carried off by a tide conceived by distant moons, taking its passengers, stunned and silent at the rail, into unknown waters.

When agricultural economists talk about the impact of the farm economy on rural communities they, too, turn to metaphors of water: economists call the process the "ripple effect." When farm prices plummet or a natural disaster wipes out a region's crops, the resulting crisis spreads throughout the area like ripples across a lake. But the geological formation known as a sinkhole holds a better analogy for what has happened to both agriculture and small towns. A ripple is, after all, innocuous, even pleasant. A sinkhole is a booby-trap.

To understand how a sinkhole is formed, imagine a large field covered with grass and the usual assortment of wild plants, bushes, small trees, and, perhaps, a herd of cows placidly chewing their cud. Beneath the grass is soil and beneath that soil is bedrock. Under this particular field, the bedrock is made of limestone, a very common rock that dissolves easily in water. As rain seeps into the ground and down through the soil, this soft rock dissolves, and slowly, a huge hole forms. The underground hole is not visible from the surface, save, perhaps, for a small opening or depression where the mantle has started to collapse. If you were standing by the edge of the field admiring the view, you'd never know that anything was amiss. There is no way to know that a few feet below the bucolic scene, the rock is being eaten away—until one day the thin layer of soil and all the grass, shrubs, herbs, and trees growing in it and the cows pleasantly grazing upon it suddenly tumble into the earth.

The period during which farmers became ever more dependent on debt financing can be compared to the sinkhole's formative years. In 1960, debt service claimed only about 10% of total farm income in America. By 1970, as increasingly expensive petrochemicals took the place of labor, that figure had climbed to 17%.[3] And in 1980, after the boom years in which farmers rapidly expanded operations in response to the exhortations of Secretary of Agriculture Earl Butz and most agricultural experts, farmers were paying out 45% of their total income to debt service.[4] Total farm debt climbed from about $50 billion in 1971 to over $200 billion in 1986.[5] The hole had grown out of sight, hidden by rising land values, a growing export market, and various government support programs. When foreign markets dried up, interest rates jumped, land values plummeted, and government funds were reduced, the thin mantle on which farmers had built their operations gave way.

The resulting devastation was quickly dubbed the "farm crisis," a term that Mike Jacobsen, a professor of social work specializing in rural issues at the University of Iowa, grew to hate. Jacobsen was deluged by reporters' calls during the few months the story was hot.

"Tell us about the farm crisis," the reporters all asked.

"Look," Jacobsen explained time and time again. "It's not really a *farm* crisis at all. It's a *rural community* crisis, and if you understand it in that way it's even scarier."

"OK," the reporters would say agreeably. "But now tell us about the farm crisis."

Jacobsen would reach for a pack of cigarettes, take one out, light it, inhale deeply, and begin again. But the message never really got out, he admits.

"Iowa is a state of small towns," Jacobsen says, leaning back in his chair at a desk overflowing with papers and ashtrays filled with cigarette butts. He speaks slowly, patiently, but with an edge of fatigue which suggests that while he hopes the scope of the general decline in rural America is communicated, experience tells him otherwise.

"These places are child-centered, clean, where you want to get married and raise your kids. The schools are pretty decent. Safe. That's the image, right? Well, the economic status of these towns is highly dependent on a number of small family farms in the surrounding area."

And today, due in part to the farm crisis, those farms are consolidating, leaving fewer and larger farms. "It doesn't take a lot of driving around to see that there's more plywood than glass in the downtown areas of these communities," says Jacobsen. "There is nothing there—these towns are just shriveling up. What you're talking about is the creation of rural ghettos."

The term "rural ghetto" is of relatively recent origin, appearing in a study of changing small towns in the Ozarks in the 1960s which traced the process toward rural ghettoization to an initial economic crisis, such as a mine or plant closing.[6] Such a severe economic jolt sets in motion three interconnected processes: (1) it begins a pattern of intergenerational poverty that families have profound difficulty breaking; (2) it touches off a class-selective migration, in which more prosperous residents move, leaving behind a community in which poverty is even more concentrated; and (3) the social and economic structure of the rural community adapts to economic shock in ways that accelerate and ultimately lock into place the downward cycle of ghettoization. Poorer communities are more likely to attract low-wage, labor-intensive industries looking for inexpensive land and a cheap labor pool. Such industries ensure the downward mobility of workers and their children. Then there is great social and economic pressure in desperate rural communities to look the other way when firms pollute the environment, mistreat workers, or otherwise exploit the area and its resources. The local supply of goods and services will diminish and so become more costly, furthering the community's decline. As the tax base erodes due to falling incomes and homeowners and industries disappear, the local government's ability to help those in need will also decline—just when the need for assistance is most critical.

The most insidious part of this process is its self-reinforcing nature; each downward step makes the next one more likely. It is for this reason that the process of rural ghettoization is so alarming.

The same process observed in a few Ozarks communities in the 1960s was at work throughout the Heartland 20 years later. The initial economic shock in this instance was the farm crisis; the sinkhole which had been growing out of sight for years didn't stop below the farmer's gate, it extended beneath the thousands of rural communities dependent on agriculture. With little income coming in, farmers' purchases were postponed—first discretionary items and then all but essentials. Inevitably, farm supply businesses were the first to feel the squeeze. Between 1979 and 1985, farm implement sales in Iowa dropped 42%. In just the first six years of the 1980s the state lost more than a third of its farm implement dealers. And in a repercussion that drives home the symbiotic relationship between farm and town, the loss rebounded back to the farmers, many of whom must now drive 50 miles or more to replace a spare part.[7]

The sinkhole quickly went beyond the farm supply businesses and nibbled away at all businesses; retail sales in Iowa's rural communities slipped an average of 25% in the 1980s. As farms began to fold, so did the many businesses that had grown up to service them. Between 1976 and 1986, Iowa's small towns suffered the following losses: the number of gas stations fell almost 41%, grocery stores 27%, building material stores 21%, variety stores 37%, men's clothing stores 38%.[8] Bankruptcies among Iowa businesses rose 46% in 1985, the largest one-year jump since records were first kept 25 years earlier.[9]

Even while these events were occurring, many predicted the sinkhole would be satiated in swallowing up only the smallest towns, communities that came to be regarded as necessary sacrifices to stem the tide. But by 1986 many of the state's 99 county seats—generally, the most prosperous communities in each county—were also suffering similar declines. On average, a town needed to have a population of over 5,000 before it showed anything but a decline in retail sales between 1976 and 1984. The magnitude of the problem becomes clear when one considers that most of Iowa's towns have fewer than 5,000 inhabitants—in fact, 479 of Iowa's 955 incorporated towns and cities have fewer than 500 citizens.

Given agriculture's growing dependence on capital over the past several decades, it's not surprising that banks became especially vulnerable to collapse in this time of economic distress.

"It's ironic because the banks that did the most to try to slow down the economic death of their communities are in the worst trouble," says sociologist Jacobsen. Those bankers who understood the dependency of farm families on town services would often stretch their investments beyond prudent practices. Says Jacobsen:

A banker will see that the town needs an implement dealer and not call his note—he'll extend his credit. And if he's doing that with the implement dealer, he's doing it with the

grocer, the hardware-store owner, a few farmers. The next
thing you know, the bank examiners come in and take a look
at the books and say, "What in the world is going on here?
You've got a debt to asset ratio of 90 to 10." And some of
those banks are taken over by a larger bank without ties to
the community and so won't make those kinds of shaky loans.
And can you blame them? In a business sense it would be
stupid to extend the loans. But what happens to the community?

In many towns, the proportion of bad loans swelled to the point
where there were more "bad loans" (those loans more than 90
days past due) than good ones. Mechanicsville Trust, for
example—the town's only bank—reported a bad-loan rate of over
65% in 1986. One bank, Citizen's State of Iowa Falls, held a troubled debt ratio (past due loans divided by capital) of a staggering
1,377% before it failed in 1986.[10]

Despite the tendency of some farmers to paint local bankers as
the villains of the farm crisis, many of these small-town financiers
were victims of the same sinkhole that swallowed up farmers and
small businesses. "Bankers, like farmers, are fighting for survival,"
declared one banker. And just like farmers, many are losing that
fight. At the peak of the farm crisis, banks fell across rural America at a rate double that experienced during the late years of the
Great Depression: an average of 67 banks failed a year from 1934
to 1939.[11] This compares to 79 bank failures in 1984, 120 in 1985,
and 138 in 1986. The recent problem has been far less severe than
during the early years of the Depression, when thousands of banks
folded—4,000 in 1933 alone. But the losses today are more concentrated in specific regions (the Midwest is the big loser) and
even more so in rural communities. About two-thirds of the modern bank failures are in rural communities, and the effects there
are dramatic.[12]

"You can't imagine what it does to a town to see its bank go
under," says Father Ronald Battiato, a Catholic priest in Verdigre,
Nebraska. The town's one bank, which held about 60% agricultural loans, was closed by the Federal Deposit Insurance Corporation (FDIC) in September 1984. Father Battiato remembers the
experience vividly:

At first we were just numb. But we didn't take it too
seriously—the deposits were insured, after all. The real trou-

ble came when the FDIC demanded instant repayment of loans. They began to confiscate any savings or checking accounts. They froze all the accounts and left many families without any living expenses whatsoever. Profits from any sales were seized and credited against loans. These people had nothing to live on. They couldn't feed livestock, repair equipment, or even keep insurance up.

By November things really got tough. We started a food pantry. I visited one family and asked how they were doing. "Oh, we're getting by," they said. I looked around their kitchen. There was no food. Absolutely nothing. And they had several small children. We took them a pickup load of food that night. I'd say we had 65 families in the area in about the same shape. It's very embarrassing for many of them—independence is a big priority out here. It angers them to take handouts, so they isolate themselves. Some families wouldn't let me deliver food during the day. I had to drive out to their place at night and drop off a load of food and leave quickly. It was pretty miserable.

For small-town banks the future is uncertain at best. While most failed banks have been quickly replaced by healthy ones, the Federal Deposit Insurance Corporation reports that it is getting harder to find banks willing to take the risk involved in such an acquisition. And as credit tightens up in the wake of bank failures, farmers and business people with anything less than a perfect credit history may find it difficult to obtain financing. As a result, local development suffers.[13]

One of the most serious consequences of the combined trends of farm and business failures coupled with bank closings has been a massive loss of jobs. According to rural development experts, for every five to seven farms that go out of business, one business in town folds. Each job lost in town means there is that much less money circulating in the community, being spent at the supermarket, the hardware store, or the barber shop. When sales drop at these other businesses, some are forced to close, and that means more jobs lost and less income to be spread around.

The sinkhole extended to manufacturing as well, an industry which was already in decline when the farm crisis hit due to the combined effects of increased foreign competition and the high value of the U.S. dollar internationally.[14] First affected were those plants making farm-related equipment. A tractor plant closed in Davenport. Dubuque lost a packing plant. Eventually, all industry in the state suffered. Job growth dried up. Iowa fell to forty-ninth place among the fifty states in job growth for the period between 1978 and 1986. The state was one of only five that saw a net loss of jobs during those years.[15]

Economist David Swenson, chief of research at the University of Iowa's Institute of Public Affairs, has made an extensive study of the effects of the farm crisis on the state's economy which "outlines a pattern of economic and social disintegration unparalleled in the country since the Great Depression."[16] Swenson found that:

> Total nonfarm employment in the state declined by 90,000 jobs (8%) between 1979 and 1983.
>
> Manufacturing jobs, generally a well-paying sector, lost nearly 59,000 jobs (23%) from 1979 to 1987.
>
> Construction lost 24,000 jobs (over 40%) between 1979 and 1984.
>
> What little growth there was in the state during these years came in the generally low-paying sectors of services, retail, finance, and insurance.

It was by confronting these trends that Congressman Richard Gephardt won the state's Democratic caucuses in February 1988. "We're trading in good jobs for not so good jobs," Gephardt warned in his stump speech—"$12 and $15-an-hour jobs for $3 and $4-an-hour jobs. The middle class is shrinking."[17]

The eventual Democratic presidential candidate, Michael Dukakis, later echoed this theme: "We're not going to accept an America where all we do is flip each others' hamburgers and take in each others' cleaning."[18] While many pundits attribute the Democrats' defeat in that year's general election to their inability to convince Americans that such a dark cloud hangs over their future, in Iowa this bleak economic scenario had already come true by the mid-1980s.

Between 1979 and 1986, the state lost 84,000 middle-class workers—those making $20,000 or more.[19] During the same period, the state saw net employment gains in the service sector of 37,000 lower-class jobs—those paying $11,500 or less.

Almost overnight, the nature of Iowa's nonfarm economy

changed. In 1979, 23% of all nonfarm jobs were in manufacturing and 18% were in services. In 1986, those numbers had been reversed: 18.7% of all jobs were now found in manufacturing and 22% were in the services sector. The effect on the typical Iowa wage earner was dramatic. That average worker was paid just over $16,000 per year in 1979. By 1986 the figure had fallen to $14,700.

"What all this shows," concludes Swenson "is a widespread devaluation of labor, an eroding quality of jobs. All net growth in the Midwest has been in the lowest paying, poverty-level jobs." Other researchers have found similar trends throughout rural America.[20]

Driving across Iowa's back roads, the observant traveler might notice plain black-and-white signs, sometimes nailed to the side of barns, sometimes to the large trees that grow directly in front of most farm houses. The drab sign reads "Century Farm." The farms may be large, with sprawling Victorian farm houses, or more modest, with one-story houses, but they are almost universally well tended. The houses are clean and painted, the lawns and fields free of weeds.

Century Farms are farms that have been in the same family for over 100 years. The annual ceremony at the Iowa State Fair in which farmers receive their signs is always well attended, and for Iowans (who by and large shun public recognition) are a rare source of chest-swelling pride. The seriousness with which the designation "Century Farm" is considered says something about the importance of family, tradition, endurance, stability—about the meaning of continuity to the people of Iowa.

And yet the rural decline of the last decade has caused a mass exodus of Iowans from their native state. Between 1980 and 1987 an alarming number of people were carried off by the state's economic collapse, like topsoil washed from a field by the rain. Nearly 200,000 more people left Iowa than moved into the state during those years—a figure that represents 6.6% of the state's total 1980 population.[21] When births and deaths for those years are figured in, the state's total population decline of 2.7% was the largest per capita loss in the nation.[22]

While Iowa is the worst example of rural population loss, the trend now extends nationwide. In one 12-month period in 1985 and 1986, rural America lost 632,000 people—the largest one-year loss in 50 years. Significant population losses were seen from New York to Appalachia to the Pacific Northwest.[23]

The population decline is taking a severe toll on rural communities, and one of the institutions hit first and hardest is the small-town school. Due to the large exodus from rural communities and the dwindling birth rate, Iowa's small-town schools have watched enrollments slide since 1969, reaching a true crisis point in the 1980s. So severe is the decline that the number of high school graduates produced by the state of Iowa is likely to fall by one-quarter by the year 2004—the second largest expected decline in the country (just behind West Virginia).[24]

The declines have touched off a debate over the issue of school consolidation. Faced with ever-fewer pupils, many argue that it makes good sense to merge schools.[25] In fact, many rural schools are already the product of mergers from past decades. Those favoring the move cite an impressive list of benefits flowing from consolidation, including savings accrued from maintaining fewer buildings and teachers and administrative staff, the ability to offer students in the larger schools more choices for courses, and higher-quality teaching due to the higher average teacher salaries possible at the larger schools.

But consolidation is not without its problems, as one angry parent pointed out in a letter to the editor of an Iowa paper:

> The largest school in our county currently has students who are picked up at 6:55 a.m. and delivered to school between 8:10 and 8:20 a.m. If my small-school math teachers taught me the correct procedure to "ciphering," that means we have students spending at least 2 and a half hours on a bus (round trip) every day of the school year (180 days). Multiplying, we find that to be 450 hours per school year. Dividing by seven hours (the approximate length of a normal school day), we find that we currently have students spending the equivalent of 64 school days riding a bus to and from school each year![26]

There is a long and continuing debate in this country over whether the quality of education at large schools is superior to that in smaller ones. While conventional wisdom has for years given the nod to larger schools, one recent study of the question declared that the trend of mergers and consolidations "may have been a very grave mistake. . . . All things being equal, smaller districts do a lot better than middle-sized districts and middle-sized districts do better than larger school districts."[27]

But all things are not equal, and Heartland small schools are indeed suffering. The 1988 "Development Report Card for the States," an annual report issued by the Corporation for Enterprise Development, gave Iowa a grade of D in the category "support for education," which includes such factors as pupil spending, pupil-teacher ratio, changes in teacher salaries, high school science and math graduation requirements, computer literacy, remedial and compensatory programs, teacher mentor programs, and teacher incentives.[28]

John Hutchinson, superintendent of the Lincoln Community School in Mechanicsville, sides with those who favor the small schools. But he acknowledges the problems caused by trends that result in high school graduating classes that can fit into a station wagon.

"We've dropped almost 300 kids since 1971–72," he says. "Out of an enrollment of 762, that's a huge loss. We've had to make a number of changes, including adjusting programs, buying supplies on a need basis only. But we offer small classes, more hands-on work with the kids. I think we offer a better education because we are small." Still, the declining enrollments have so concerned the local school board that it has taken the unprecedented step of advertising the benefits of Lincoln Community School District in area papers, hoping to entice students into the school.

Rural districts are contemplating using similar tactics to draw high-quality teachers to rural areas. Saying that rural educators face a task of "Herculean proportions," one United States Department of Education official suggested that the state provide incentives such as subsidized housing and loan forgiveness programs to teachers who agree to teach in rural communities for a specific length of time. The exodus of Iowa's young couples and their children has Iowa educators and politicians scrambling for other Herculean solutions. These range from linking schools by satellite, microwave, and fiber optic technologies to using school buildings as day-care and senior-citizen centers in addition to their traditional use.

Despite these new proposals and trial programs, education officials facing a continually shrinking student population continue to turn to consolidation as the solution of choice. That solution, however, can devastate already-hurting small towns. When two schools merge, one is forced to close. While this often results in a slight increase in staff at the remaining school—sometimes the move results only in larger classes, however—for the town losing its school, consolidation inevitably means the loss of several jobs. Indeed, a net payroll reduction for the two schools is often listed as one of the primary benefits of consolidation. For the many small towns in which the school is the major employer, the process represents a major economic setback.

The effect of consolidation on a town's pocketbook is slight, however, compared to the damage the process does to the heart of a community, for the red brick schoolhouse sitting squarely at the center of most small towns is also found at the crux of community life. From the local high school football and basketball games, to the annual kindergarten-through-fifth-grade Christmas concert, small-town life revolves around the many school events which mark the seasons for townspeople in the same way the perennial cycle of planting, cultivating, and harvesting does for farmers. In many towns, these school activities are the only regular social events—for adults as well as for their children. After consolidation, many parents are unwilling or unable to make the long drives involved in attending these events.

Residents often miss most the smaller daily events, the ones that, woven together, create the unique texture of small-town life—events that pass unnoticed until they are gone, such as hanging wash out on the line and hearing the distant shouts of children playing at recess. A town devoid of children is a town devoid of hope. Many parents fear that attending school in a different town will loosen the bonds that join their children to the community, making it more likely that someday those children will leave town for good.[29]

The dissolving ties between individuals and their communities is manifest in another trend that bodes ill for the future: a "brain drain," a trend among college graduates to leave Iowa once they have their degrees. Only one-third of seniors at the University of Iowa plan on remaining in Iowa after graduating.[30] Worse yet, 87% of seniors majoring in engineering and 74% of those majoring in business—areas vital to the state's plan to expand into high-growth fields—say they intend to leave the state.[31] This intent has translated quite readily into fact: four of ten Iowa college graduates actually left the state after receiving a diploma in 1986.[32] A majority of those who left also said they would have stayed had a job been available. Once they're gone, the investment made in the education of these young people is forever lost to the state.

Take a stroll around the town square in almost any small town in Iowa and one thing will strike you immediately—the age of the people you meet. A surprising number of the residents are elderly. The sharp increase in the average age of Iowans is a result of the state's declining birth rate and increasing outmigration of young families. The large proportion of elderly citizens here is expected to climb.

"By the year 2010," the state's Department of Management predicts, "there will be an explosion of the elderly population."[33] Some say that explosion has already started. The average age in many Iowa towns is over 50 years. The percentage of residents over the age of 65 in the state is exceeded by only three other states, two of which—Arizona and Florida—are retirement centers.[34] The problem is not limited to Iowa, but is common throughout rural America. In some rural Nebraska counties, for example, more than 30% of the population is over 60 years old.[35]

For many small towns, this "graying" process could be fatal, for changes in age composition are the most important factors in determining whether a community will continue to exist.[36] This is true, simply because once a town has a high number of elderly residents and a relatively low number of childbearing families, it can no longer maintain its population. The situation is less of a problem for states such as Florida or Arizona, which can maintain their populations thanks to the retirees who are constantly moving there from other states. Iowa's situation is very different, however.

Aside from the possibility of the community's actually ceasing to exist, the graying process poses a number of other serious challenges to small towns. How, for example, does the state provide adequate health care for the increasing elderly population at a time of diminishing revenues?

"Medicare simply does not cover very much of the cost of home health care that our clients need," says Aileen Holthaus, a field organizer for the Iowa State Council of Senior Citizens. "It also doesn't cover prescription drugs, dental care, hearing aids, nursing-home care, eyeglasses—the kinds of things the elderly really need. Many of the people I see pay around $100 a month for prescription medicine. A lot of them used to depend on their children and grandchildren for help, but they've moved out of the state looking for work, leaving a great many older Iowans to fend for themselves."

According to council president Jack Seeber, "The rural elderly are at a disadvantage. In most cities in Iowa we have programs for the elderly, including meals on wheels, companion and transportation programs. But these things just aren't available in rural areas. There are a lot of people hurting out there."

For these elderly occupants of rural ghettos, leaving isn't an option: they are ill-prepared physically, emotionally, and economically to pick up and start a new life somewhere else at a time of life when, in the American ideal, they are supposed to be enjoying the fruits of a lifetime of work. Many of them are retired farmers, and most are living on Social Security benefits. The only equity they have is in their homes, for which there is no market.

"So many of our old people are confused and frightened by all the changes," confided one small-town minister. "They're down to one doctor, one grocery store, and the question they keep asking is, What happens when the last doctor is gone, when the last grocery store closes? And how do you answer that? What in God's name are they going to do?"

In the summer of 1988, the same sinkhole that had already consumed farmers, small businesses, young families, and schools began to nibble away at the edges of another important rural institution: the small-town hospital. The residents of Hamburg, Iowa, population 1,600, woke up one morning to find that the hospital on which they had depended for health care for some 65 years, in which many of them had, in fact, been born—the Grape Community Hospital—had filed for bankruptcy.

"People were shocked," says John Field, editor of the local weekly paper, the *Hamburg Reporter*. "We've just always had a hospital in Hamburg—it defines the town. There's hardly anyone around here who can remember when we didn't have one."

Marvin Vollertsen, owner of the town's only pharmacy, was one of a handful of Hamburg residents not surprised by the news. He had known that the small hospital had been in trouble for some time, partly due to poor management at Grape Community. But the hospital had also fallen victim to a larger trend.

"We may have been the first to file bankruptcy, but we won't be the last," warns Vollertsen. "All these rural hospitals are in trouble."

They are indeed in grave trouble. According to Dr. Stephen Wright, director of rural health at Georgia Southern College, rural America can be described as a "health disaster area."[37] If rural hospitals were human patients, doctors would place most of them in intensive care. At least 20 of Iowa's rural hospitals may not survive another decade, according to one expert, and those that do remain will have to dramatically change their mode of operation.[38]

The Grape Community Hospital (GCH), like similar institutions across Iowa, has been hard hit by the state's population decline. While GCH is licensed for 49 beds, it averaged only 8 patients per night in 1988. On some nights there were as few as 4 patients. Small-town hospitals, including GCH, were also severely hurt in 1981 when Medicare set reimbursement rates for rural hospitals almost 40% below those allowed for urban hospitals.[39] Since rural hospitals depend far more upon Medicare payments for their incomes than do their city counterparts, the situation is especially devastating. Coupled with the decline in patient use and an increase in the number of uninsured patients, the cut in government funding has left many hospitals in desperate financial straits. In a belt-tightening move, many hospitals have trimmed staff to a bare minimum. At one hospital, for example, nurses are now forced to do double duty as ambulance drivers.

While particularly severe in Iowa, the problem is widespread across rural America. Montana Senator John Melcher said in 1988 that 161 rural hospitals had closed since 1980, and that at least 600 of the remaining 2,700 rural facilities were "on the brink of closure."[40]

The closing of a small-town hospital, just like the closing of a school or the loss of several farms, touches the lives of community residents in many ways. Most immediately, it means a decline in the availability of health care. As the *Des Moines Register* pointed out in an editorial, rural America "has a special need for emergency care because of the age of rural residents and the high risk of accidents in farming. When the patient is in shock, Omaha or Kansas City can seem half a world away." Already, compared to the rest of the nation, Iowa ranks close to the bottom (forty-eighth) of the list in terms of money provided for emergency medical services.[41]

The closing of a hospital also means the probable loss of many local physicians, who may move to be closer to a large medical facility. And it will certainly make it difficult for a rural community to attract doctors to replace those who retire or move. Rural Iowans, as rural people throughout the country, already have far less access to health care than do urban residents—despite the fact that the per capita need for medical care is greater in rural areas. Fewer and fewer doctors have settled in rural areas over the past several years, due to a combination of factors including lower rural fees, higher malpractice insurance costs, a paucity of modern medical equipment, and a general lack of amenities. In Iowa's most urban counties, citizens enjoy a resident to physician ratio of 270 to 1. In the most rural counties that ratio widens to 1,500 to 1.[42]

But the loss goes beyond access to health care.

"If the hospital closes up," says Hamburg's Vollertsen, who has worked in Stoner Drugs for over 42 years (the store has been open since 1896), "my business would have to be scaled down. If GCH closes . . . well, I don't even want to think about what will happen if it closes." Vollertsen supplies prescriptions for the hospital on a part-time basis, usually two or three days each week, a relationship that will end if the hospital folds. In addition, the hospital employs between 60 and 70 Hamburg residents, all of whom may soon be out of a job. For all these problems, Hamburg is more fortunate than most communities facing the loss of a hospital. The town has two other major employers to take up some of the slack that the anticipated layoffs will produce.

Despite its obvious importance, health care is just one of many basic services that have been undermined by the widening sinkhole, services most Americans take for granted. The ability of local, county, and state governments across America's Heartland to simply govern has been called into question.

In the 1980s rural communities were hit by a double whammy. First, local tax revenues—the stuff that local services are made of—followed personal incomes and land values down into the sinkhole. Ordinarily, the local governments, finding their coffers empty at a time of increased public need, could turn to the federal government for help. But the early years of the decade witnessed the opening salvos in the "Reagan Revolution's" fight to "get government off the backs of the people." This revolution proved most successful, however, at knocking rural communities off their feet and flat on their backs. From 1980 to 1985, as farm income plummeted throughout the Midwest, the federal government aid so desperately needed by state and local governments was cut by one-quarter.[43] The reduction in tax revenues and federal aid has meant cutbacks in virtually every area of local service, as everything from public libraries to hospitals, from police forces to town softball teams, has slipped down into the sinkhole. Like the hand-lettered signs throughout the Schultz Brothers Variety Store during its final days ("No Lay-Aways, No Returns, No Checks"), the signs of decline confront rural residents everywhere they turn.

There is a final, bitter irony for many small-towners: the more their towns adapt to decline "successfully," the more they ensure further decline. When services are cut back to keep out of the red, the community's "quality of life," that hard-to-measure but all-important value, drops. And as it drops, so does the population. People who can move to a more amenable location will do it. Those who are left are generally the poorer rural residents—those more dependent on county, state, and federal programs. They are also less able to contribute the tax dollars needed to fund services, and so the downward spiral of pain and poverty continues.

While the original study of rural ghettos in the Ozarks determined that an initial economic jolt—such as a farm crisis—was needed to initiate the process, there is good reason to doubt that this is so. In a study prepared for the government in the early 1940s, Walter Goldschmidt compared the quality of life in two California towns, Arvin and Dinuba, investigating the link between rural decline and changes in the structure of agriculture.[44] The towns, which were similar in population, differed in one important way: Arvin was surrounded by large-scale "industrialized" farms, Dinuba by small "family-type" farms. Goldschmidt's findings were striking: Dinuba supported twice as many businesses as did Arvin, with 61% greater retail volume. Dinuba residents enjoyed more parks, schools, churches, newspapers, public recreation centers, and civic organizations. Dinuba residents also had a higher standard of living than did residents of Arvin, and in addition had more control over community decisions: those decisions were more commonly made by local popular election in Dinuba and by county officials in Arvin. Greater social stratification was found in the community of large farms, and greater economic and social homogeneity of the entire population existed in the town of small farms.

Revealing the short-and long-term negative effects of an industrialized farm system on rural communities, the study prepared decades before the farm crisis of the 1980s suggests that rural ghettos are not necessarily dependent on an economic crisis for their creation. Rather, the structure and scale of agriculture itself determine the health and viability of a community and its inhabitants. The implications for the future are clear and sobering: though the recent financial crunch called the farm crisis may be tapering off, the rise of the rural ghetto is likely to continue as long as industrialized farms eliminate smaller ones.

Social critics use the term "marginalization" to refer to those individuals who have been pushed beyond the edges of society's consideration—people who are poor, the homeless: people who no longer count. These people are residents of the United States, but they are no longer citizens in any meaningful sense of the word. In rural America, we are now making the terrible (and yet wholly logical) leap from marginalizing individuals to marginalizing whole communities, and perhaps even to rendering an entire region superfluous to the flow of American life. As politicians profess a deep and abiding love of "Heartland values," thousands of small towns spread out across the American countryside are left to wither on the vine.

4

Poverty and
Social Disintegration

*We won't be happy until the American recovery stretches
across this country like a blanket with the Midwest safe
and warm inside.*

RONALD REAGAN
in a Campaign Speech, 1984

Highway 30, Iowa

To the occupants of cars speeding down the stretch of open
road between the towns of Calamus and Ground Mound he
barely exists: a momentary blur against the empty early-winter
fields that flank the road on either side. For his part, he is as
oblivious to the passing cars as they are to him. He walks slowly
but steadily through the ditch beside the road; bent at the waist
like a tipped-over L, he peers through thick lensed glasses at
the ground in front of him. From time to time he stops,
reaches into a tumble of weeds, and withdraws an aluminum
can which he shakes dry and drops into the white plastic garbage bag he carries over one shoulder.

Every day, weather permitting, he walks these roads collecting the cans which bring a nickel each at a local convenience
store. At 69, his pace is slow, but he can cover at least eight
miles a day—a distance which translates into $10 to $12 in cans.
Even with weakening eyesight, he is expert at distinguishing the
valuable from the worthless in the trash that litters the Iowa
roads.

"Those little cellophane bags can be buggers, though," he admits in a reedy voice that is often lost in the wind which blows steady and hard from the northwest. "When the light hits them right, they can trick you. My knee doesn't work too good any more so I poke around with my foot before I bend over. I still get fooled sometimes."

But he isn't fooled by the seeming worthlessness of ordinary brown paper bags. They can be a gold mine, he says, because they often hide a six-pack tossed out the window by a teenager who panicked when what looked like a sheriff's car appeared out of nowhere in the rear-view mirror.

He is no stranger to hard times. He lived through the dust bowl on a farm in Kansas and then moved to Iowa and scraped together the cash to buy a 35-acre farm. He sold the land years ago and moved to town, and when times got tough he went to work in a nearby family-owned business. In 1985, when the farm economy went completely bust and the company laid off anyone who wasn't immediate family, he took to scouring the roads that crisscross the flat treeless land just west of the Mississippi, in search of cans. His children moved away, a daughter to a good job out east, a son to Des Moines, where he is still looking for work.

"But what good does complaining do?" the old man says. "You do what you have to to get by, that's all. Say, be careful picking those things up," he adds, deadpan, when I hand him a can. "You might catch AIDS."

I ask if he really believes that.

"Naw," he says and smiles. "I just tell people that to cut down on the competition."

There *is* growing competition for cans along the Iowa roads. In a given 50-mile stretch, you are likely to see one or two gleaners wading through the weeds, plastic bags thrown over one shoulder. Sometimes, especially in the summer, you will come over the crest of a hill and encounter an entire family collecting cans together, their battered car parked off to the side of the road. But that is still rare. Can-hunting is more usually a solitary activity, done by farmers and laid-off factory workers while the kids are at school.

"It's the same old story," the old man says as he kicks with his good leg at what turns out to be an empty Doritos bag. "The rich create these hard times, and they're doing OK. It's the little guy who gets hurt. And things are going to get worse, you know. A lot worse. People think they've got it tough now. Hah! Just wait."

He kicks again at something shining up from beneath a clump of dry brown grass. This time he is in luck—it is a can, crushed almost paper-thin under the wheels of a car or truck. The old man picks at the dirt and gravel that fill its many crevices and then holds the scrap of metal outstretched in front of him. He smiles, regarding his find for a moment before gently placing the crushed can into his bag, and—resembling a modern-day Johnny Appleseed—continues down the road.

The old man, in his age, his poverty, and his wonderful ability—and awful willingness—to endure hard times stoically, is as nearly perfect a symbol of late-twentieth-century rural America as one could hope to find. But most of all he embodies the plight of small towns across the country in his near invisibility.

To most Americans, rural communities are just dim blurs alongside the gleaming superhighway that carries us into what we tell ourselves is an ever-brighter future. If we notice those blurs at all, it is usually to laugh at their quaintness, perhaps warmly à la writer and humorist Garrison Keillor, or to shake our heads at the backwardness of our unfortunate rural cousins. Few, however, slow down enough to allow the blurs to differentiate themselves into real people, in real communities, with real problems to be solved or ignored. Time has not forgotten Keillor's Lake Wobegon; we have.

This kind of ignorance about the problems of rural America is astonishing, especially in an era referred to, reverentially, as the "information age." Thanks to the computer revolution, we now have the ability to gather and analyze previously undreamed-of amounts of information. And thanks to the same technological miracles, that information is also more accessible than ever before to politicians, government agencies, journalists, and the public.

We can calculate with amazing accuracy, for example, how many refrigerators in a given rural zip code contain Hellmann's Mayonnaise and how many instead contain Kraft Miracle Whip salad dressing. And yet, somehow, we can't say how many children in that same area go to sleep each night, not in the comfort and safety of their own beds, but in a public shelter, or in a tent, or in the back seat of the family automobile.

This ignorance amid an ocean of information tells us something about ourselves. Of course it says the obvious: that we don't like to dwell upon the suffering of others. Who does? But it also suggests something about our view of rurality.

We like to think of small-town America as being impervious to the grim trends of poverty and social disintegration that are rampant elsewhere. It gives us a certain measure of confidence to believe that whatever problems are wracking our cities, whatever demons of entropy are dismantling our cherished institutions of family, hearth, and home, the nation's small towns somehow will stand forever as a bulwark against these modern failings.

This idealized picture of small-town America is our security blanket, and no matter how ragged that blanket is getting around the edges, we fiercely cling to it as we dream our American dreams. Even the most cynical among us—even those who sneered at Ronald Reagan's assertion in 1984 that "it's morning in America"—seem inclined to believe that it is morning in America's small towns, and will remain so forever.

The truth is very different, of course, for the changes sweeping through rural America suggest that it is really twilight, not morning—at least for a great many small towns.

Rural poverty is not, however, a new problem. (Unfortunately, that observation, which should be the starting point in the discussion, usually signals the end of debate on the matter. Many seem to hold the perverse belief that because the problem is long-standing, there is less cause to attend to it.) Our nation's experience with rural poverty stretches far into our past. For example, the flood of impoverished Southern tenant farmers searching for a better life "out west" in the 1890s became so great that many simply wrote the initials G.T.T. across their boarded-up front doors as they departed. Everyone passing by knew what those letters stood for: Gone To Texas.

But in each generation we seem condemned to "discover" rural poverty anew. A book published in 1927 declared, "Rural life in America is decaying. It is slowly slipping down toward peasantry."[1] Just 12 years later Americans were shocked by James Agee's and Walker Evans's chronicle of Southern poverty during the Great Depression, *Let Us Now Praise Famous Men*.[2]

The most comprehensive modern governmental study of the problem was undertaken in 1967 by President Johnson's National Advisory Commission on Rural Poverty. The resulting report, *The People Left Behind*, documented the many effects of poverty, from malnutrition and actual starvation faced by the children of rural America to the lack of housing, medical care, education, and political power of the rural poor. The report concluded with a cornucopia of recommendations to foster an ambitious goal: "The Commission is convinced that the abolition of rural poverty in the United States, perhaps for the first time in any nation, is completely feasible."[3]

It is doubtful that rural poverty could have been completely wiped out had the commission's recommendations been fully implemented. Certainly the recommendations (which included both short-term emergency aid, such as food assistance programs, and long-term structural changes in education and rural development) would have gone a long way toward reaching that goal. But they were never given that chance. "The recommendations weren't implemented at all," says C. E. Bishop, executive director of the commission and now president emeritus of Houston University. "We got into the Vietnam War and that ended that." The war effort captured the country's attention and sucked its resources dry; America's leaders had to choose between bombs and butter, and as often happens, the bombs won out. The Johnson Administration thanked Bishop and the commission for their excellent work and then promptly buried the report.

Even so, the quality of life in much of rural America rose during the 1960s and 1970s, thanks, in large part, to the institution of general, as opposed to rural-only, federal income security and assistance programs. Food stamps, Supplemental Security Income, and Aid to Families with Dependent Children (AFDC), as well as the expansion of Social Security benefits, helped to narrow the gap that had separated the two Americas, urban and rural, almost from our nation's earliest days. But a crushing poverty is the most durable of all rural institutions. Even in the "good" years between 1960 and 1980 (when the rural poverty rate was cut in half) a significant number of rural communities remained mired in poverty.[4] In fact, the majority of the nation's poorest rural counties have been poor for over two decades.[5] Today, while many of these traditionally poor rural areas remain impoverished, rural ghettos are also taking root in previously prosperous Heartland communities. A growing number of rural Americans today find themselves—in the words of the 1967 report—"on the outside looking in."

The generally prosperous picture most Americans have of the nation's countryside merely serves to mask widespread rural poverty.[6] Of the 54 million people living in rural America today, over 9 million exist below the poverty line, a level of poverty that is nearly as high as in the nation's blighted inner-city neighborhoods.[7]

Although rural poverty strikes farmers disproportionately (one-third of all farm families are poor),[8] poverty has spread far beyond the farm gate. Today farmers and their families comprise a surprisingly small minority of the rural poor population—only about 10% of the total.[9] In fact, the composition of the rural poor is very similar to that of the urban poor. For example, the groups most vulnerable to poverty in the nation's cities—the elderly, minorities, single mothers—are just as much at risk when they live in rural ghettos.[10] Children are even more vulnerable in rural areas: one out of every four rural children today is poor, while about one of five urban children lives in poverty.[11]

Unemployed workers also make up a significant share of the rural poor population, a fact that challenges popular stereotypes. When most Americans think about the country's unemployed, the picture that comes to mind is one of city dwellers, those in the inner cities who spend their days hanging out on the streets getting into or trying to keep out of trouble. For many years that image had some truth to it: cities had a higher rate of unemployment than nonmetropolitan areas. However, in 1980 the situation was reversed, and the gap has grown every year since.[12]

But simple unemployment figures do not measure the true level of economic hardship in rural America. A better method is to break unemployment into its components, including discouraged workers who are no longer looking for work and involuntary part-time workers. In these two categories, rural people suffer more than urban residents: compared to their city counterparts, more rural adults have given up hope of finding work and more work inadequate hours.[13]

The rural poor differ from the urban poor in an even more important facet of unemployment: their ranks contain a far higher percentage of workers. Even when working full-time, year-round, the residents of rural ghettos are far more likely to remain trapped in poverty than are urban workers, due to low wages.

Donald Webster, a 34-year-old truck driver in eastern Iowa, is typical of the new rural ghetto resident. The father of two children under the age of ten, Webster has worked for the same trucking company for the last five years. Although he works full-time—often on the road for nine weeks at a time—he does not bring home enough money to raise his family above the poverty line. The Websters' daughter, Brenda, was born with a serious respiratory problem. Because Donald's job does not provide medical insurance, Brenda's last hospital stay depleted what little savings the family had managed to set aside. The extra medical expenses added to the cost of feeding and clothing a family of four caused the family to fall behind on payments for their tiny two-bedroom house, and they moved into a small trailer on the edge of town.

"Our choice this month," says Donald's wife, Donna, on a bitter cold January morning "is whether we should buy enough heating oil so Brenda doesn't get sick in the first place, or save the money for medicine in case she does."

"I'm doing the best I can, you know?" adds Donald. "But it doesn't seem to do any good."

For an increasingly large number of rural Americans just like the Websters, work no longer provides a way to escape poverty. The vast majority of able-bodied rural poor family heads worked at least part of the year in 1987; just below one-quarter of them worked full-time, year-round.[14] Despite their efforts, the low wages they were paid were simply inadequate to lift them out of poverty. More than two out of every five rural workers (42.1%) earn a wage insufficient to lift a family of four out of poverty even when working full-time, year-round.[15] A decade ago, this figure was 31.9%.[16]

What is particularly discouraging about the current high level of rural poverty is that its growth in the 1980s wiped out a decade of steady improvements won in the 1970s. The rural poverty rate dropped from 16.9% in 1970, to 15.4% in 1975, down to 13.8% in 1979. It then peaked at 18.3% in 1983, and in 1987—the last year for which statistics are currently available—dropped to 16.9%, precisely where it stood in 1970.[17]

The people of Iowa, once a mostly middle-class state touted as a "land of milk and honey," have fallen father and faster into the sinkhole of poverty than have the residents of any other state in the nation. Between 1979 and 1985 the percentage of Iowans living in poverty more than doubled.[18] In those few years, Iowa's poverty rate went from 7.9% (comparable to that in New Hampshire, Rhode Island, and Colorado) to 18% (almost identical to the rate in Tennessee, Louisiana, and Georgia). In 1985, nearly one out of six Iowa residents was living below the federal poverty line.

The spread of poverty touched off by the farm crisis has been felt throughout the state. "Almost everyone in rural Iowa has been affected by the farm crisis," says Lloyd Gehring, operations director for a local private nonprofit community action program in Iowa. "Rural America is just suffering. We're all tightening our belts." Even the figure showing Iowa with an 18% poverty rate may be misleadingly optimistic. The state percentage is skewed by the relative affluence of its more urban counties. Examining poverty statistics on a county-by-county basis, it quickly becomes apparent that poverty is concentrated in the state's small towns; some rural Iowa counties have poverty rates of 30 percent.[19] And yet, because of the nature of rural life—isolated, dispersed, out of public sight—we are not aware of its extent.

The increase in rural poverty has brought a host of other social ills which run counter to our traditional image of Heartland America: problems such as hunger, malnutrition, and homelessness. A recent report on rural poverty by the group Public Voice for Food and Health Policy reads like a description of a Third World country.[20] The researchers found growth retardation among children, infant mortality rates 50% higher in rural areas than in urban ones, an increase in the number of low-birth-weight infants in rural counties, nutrient deficiencies such as iron anemia, and diets dangerously low in such important vitamins as A and C. The group concluded that if the infant mortality rate in the nation's poor rural counties had been as low as the national average, the lives of between 4,000 and 6,400 babies could have been saved.[21]

Overall, the reports confirmed and expanded on an earlier study by the Harvard School of Public Health's Physician Task Force on Hunger in America. That study documented widespread hunger across the country, including Iowa.

Waterloo, Iowa

It is the kind of quiet desperation we associate with other places. Entire families traveling from town to town to look for food or a job. Grown men foraging through junk yards to find items of value. Children coming on their own to feeding centers in hopes of a meal.

But it's Blackhawk County, Iowa, right in the heart of America's breadbasket. "It's awful," a local grocer reported.

"This morning a man stood in the checkout line with a loaf of white bread, powdered milk and two cans of dog food. I looked him in the eye and he turned red and looked away."

It startles the visitor to drive into Waterloo, surrounded by fields of corn, to find a van with the words "Food Bank" painted on its side. But feeding the hungry has become serious business here.

"Last summer we offered a feeding program for the town children," explained a local school official. "We expected 300 children, but more than 2,000 came the first day. We thought we knew our community but we were in for a big surprise."[22]

The very idea of hunger in rural areas, and especially in the Midwest, surprises many, and for obvious reasons. It is simply hard to reconcile malnutrition with an area that boasts the world's most fertile land. But of the country's 150 worst "Hunger Counties" (as identified by the Harvard group), 97% are in rural areas.[23]

One indication of the depth of the problem was seen when the government announced plans to scale back its emergency food distribution program in Iowa in May of 1988. The program, under which cheese and other dairy products were given out to needy Americans, was begun in 1982 mostly as a means of reducing government-owned stocks of those commodities. Rather than dump the food, the government decided it made more sense to give it away to those who needed it. By 1988 the stockpiled food was reduced to a level deemed reasonable by government analysts, and the program was to have been cut. The routine announcement was greeted by a public outcry. The head of Iowa's distribution program said that any cutback would be "devastating" to the hundreds of thousands of Iowans—more than 10% of the state's population—who depended on the monthly allotment of cheese, butter, and powdered milk.

Another indication of the magnitude of the problem: recently at a private food distribution center in rural Missouri, people waited in a line of pickup trucks so long that it took four hours to reach the head, just to receive a single box of food.

"If that many of these proud, fiercely independent people are willing to be seen in public in a line waiting for food, they must be awfully hungry," commented a professor of social work at the nearby University of Missouri.

The problem of hunger and malnutrition is especially severe for the children of the rural poor, who are far more likely to have inadequate diets than are the children of the urban poor. In fact, it was recently determined that "inadequacy of fruit and vegetable consumption among the rural poor is the most distinguishing dietary characteristic between the rural poor and the more urbanized impoverished population."[24] In other words, the children of the rural poor, who grow up amidst this country's amber waves of grain, are *least* likely to have a proper diet. And worse; it is the youngest of these children, those between two and five years old, who suffer most from malnutrition.[25]

The children of the rural ghetto generally start the day drinking an overdiluted mixture of Tang or Kool-Aid instead of orange juice. For many of them, the school lunch is the only hot meal of the day. In the evening, these children sit down to meals that are wholly inadequate to the needs of their growing bodies, and so they are prey to a variety of nutritionally linked diseases such as anemia and chronic intestinal parasite infestation. But the damage done by these nutritional deficiencies goes beyond purely physical development—the mental well-being of rural children of all ages is also at risk. Infants who suffer from a deficiency in iron score below iron-healthy babies in tests of mental development. Iron-deficient preschoolers have difficulty concentrating, and older schoolchildren who are low in iron score lower in educational achievement tests than do healthy children In an age in which educational achievement plays an increasingly important role in determining who will succeed and who will not, the nutritional deficiencies suffered by the children of our rural ghettos cast a shadow over their future.[26]

Many urban people are surprised at the extent of rural hunger. Why, they ask, don't the rural poor simply grow the food they need? That would seem to be the obvious answer for hungry people living in America's breadbasket, but for a variety of reasons it is not a practical solution. First, a large number of the rural poor are elderly and cannot maintain a garden large enough to supply their food. But what about those who are young enough? After all, traditionally in the Midwest, the woman did raise a garden while her husband was farming or working at some other job. But during the economic downturn of the last decade both parents have been forced to go to work. Having one adult stay home is today as much a luxury for rural Americans as it is elsewhere. As a result, most rural families simply do not have the time needed to grow their own food. And those with the time don't have the money. A garden large enough to feed a family requires a sizable cash investment for seed, fertilizer, and equipment. It is also a risky investment. If there is a drought or if a blight wipes out a crop, the family ends up with neither home-grown produce nor the money necessary to buy food.

The government's response to rural hunger can be charitably termed "inadequate." Even compared with federal programs for the urban poor—certainly not our best-served population—what's done for the rural poor doesn't measure up. In program after program, from food stamps to general assistance, from school lunch to summer feeding programs, the rural poor are consistently underserved when compared to the urban poor, and they are underserved in growing numbers. From 1979 to 1983, for example, the number of rural poor who were eligible for food stamps but who did not receive them swelled from less than 5.7 million to 7.5 million. Though the rural people make up 30% of the nation's poor population, they receive only 20% of federal government poverty funds.[27]

A large part of the reason for this disparity is due to the different compositions of the two populations, rural and urban. Because the rural poor are far more likely to live in two-parent families than are the urban poor, a greater percentage of them are ineligible for programs such as Aid to Families with Dependent Children.[28] Farm families are even less likely to qualify for federal antipoverty programs because, even when they lack the money for basic necessities, they *appear* well-off on paper, due to ownership of farm equipment. Hot lunch programs—among the most important federal programs for rural children—also have a built-in bias against the rural poor. Rural poor children are only half as likely to participate in summer lunch programs as children of the urban poor, in large part because of recent regulatory changes that prohibit private nonprofit organizations from administering these programs. While only a few such groups sponsored summer meals in the nation's cities (schools more often do the job there), rural areas depended on the participation of church and social groups. In some states, up to one-half of the rural children who had received summer lunches became ineligible because of the restrictions.[29]

Perhaps even more shocking than the existence of hunger

amidst the plenty of the nation's breadbasket is the rise of homelessness throughout small-town America. Unfortunately, we can
only guess at the true dimensions of rural homelessness because
there has been no comprehensive, nationwide study of the problem. With a few exceptions, the subject has been ignored. According to most "opinion leaders" and policy makers, there simply is
no problem of rural homelessness: a lack of adequate shelter is
pronounced an exclusively urban problem. Once again, the rural
poor are the most invisible Americans—unseen by even those
groups set up to be advocates for the poor.

When we do talk about rural homelessness, we usually focus on
only the most visible and most easily understood part of the problem: those people who are living on the streets. Even under this
very limited definition, rural America has a problem. One 1987
study reported a "sizeable and steadily growing population that
lives under bridges, in cars or abandoned buildings, in emergency
shelters" throughout the rural Midwest.[30] The number of nights
spent in a shelter for all homeless Iowans swelled from 28,000 in
1984 to over four and a half million just three years later.[31]

But the very real problem of rural people living on the streets
or in shelters is just one small part of the picture, according to
experts like Calvin Streeter, a sociologist at the University of Missouri. "You have those people who lack any kind of permanent
shelter," he explains, "and then you have those who have a form
of shelter, but that shelter is not considered their home." In rural
America, the overwhelming majority of the rural homeless exist
in a limbo called doubling up. Unable to pay rent or make house
payments, rural people move in with relatives or family for short
periods, and then, when they've worn out their welcome, they
move on to another house.

"I think we are still denying we have a problem with homelessness," says R. Dean Wright, author of a recent study of the problem. Wright found that nearly 15,000 Iowans are homeless and
another 37,000 live on the edge of homelessness. He also found
that more than one-quarter of that total are children.[32]

"If we erred in our report, we erred on the side of conserva-
tism," Wright says. "Iowa is one of the worst-case scenarios. People throw $1 million here and $1 million there and they say 'We've
cured the problem.' But we haven't. We have a nice shiny new
shelter in Des Moines, but the problem is bigger than that, and it's
getting worse."

Father Frank Cordero, a Catholic priest who has long been active in rural issues, calls rural homelessness an invisible problem. "We hide poverty much more easily out here than in the cities," he says. "Homelessness in Iowa's cities is growing, but it's just the tip of the iceberg—they're the people who have fallen completely through the cracks—[there are] thousands more in rural areas who are unemployed or underemployed, living in inadequate housing with no health care, just barely making ends meet and being forced to choose between spending for food or housing."[33]

The story Bill Howard tells provides a stark picture of life for the rural homeless. Howard is a 41-year-old Vietnam veteran who was forced to move in with his mother after losing his job and falling behind on his own house payments. The furnace in his mother's house was broken when we talked, and the two didn't have the money to fix it. It was the week before Thanksgiving, and the temperature had fallen to 20 degrees outside. Inside the house the mercury hovered in the mid-30s.

> You're always on the verge of just losing everything. Even though you may temporarily have a roof over your head, there's always the threat of being out on the street. I almost reached the point of paralysis, because of the fear of not knowing what I was going to be doing two or three days down the line—whether I was going to be living on a street corner and simply freezing to death. It made it almost impossible for me to function.
>
> You condition yourself to accepting whatever you can get. If the living conditions are shabby, well, you know, that's my place in life. You deal with it until the anger and frustration take over. The only alternative you see is to move on, to go someplace else. But what kind of adult life are these kids who are always moving from one crummy place to another going to have? Every year or two they're moving, moving, moving. They're going to be conditioned to the same sorts of things their parents were. They'll never have any roots. It's a vicious circle. It's never going to get better. It's just constantly going to be like a whirlpool pulling you down and down and down and down.

As the late author Michael Harrington pointed out, there is more to homelessness than simply lacking a roof over one's head.

A home, wrote Harrington, "is the center of a web of human relationships."[34] In rural America, that web of human relationships—the extended family and the community—has been particularly important in defining people's lives. But as the ranks of the poor grow, the proud tradition of "taking care of our own" is flickering out: It is difficult enough to take care of *yourself* in hard times. More and more rural good Samaritans—those who have taken others into their homes—are finding themselves descending into poverty because of the increased burden assumed in doubling up.

Family members who depended on each other are dispersed like atoms—one to track down a rumored job over in Des Moines, another off to the Sun Belt, where things must surely be better. Unemployed men in search of work often leave their wives and children with relatives or friends. Sometimes things work out, jobs come through, the family gets back on its feet, and they are reunited. But frequently the strain of separation and poverty tears families permanently apart.[35] The new migrants wander between cities and then wander within the cities, and many of them will never come to rest, to take root within a new environment and flourish. They will merely change, as one observer put it, "from being rural poor to being urban poor."

The transition to city life is difficult for rural people with limited education and possessing little or no money. According to Reggie Goldberg, director of Jewish Family and Children Services outside of Kansas City, their problems are similar to those faced by refugees coming from other countries.

"It is a tremendously difficult transition for many of them," she says. "They are so afraid in the city that some of them sleep with guns under their pillows. And their children face discrimination because their clothes and values are different."[36]

But hard as life is for those who migrate to urban areas, it is at least as hard for those who are left behind in the nation's rural ghettos—out of reach of most welfare programs and even farther out of the public eye than are their urban counterparts. According to R. Dean Wright's study of homelessness in Iowa, probably the most severely affected by these trends are elderly women. Says Wright, "We found that there's a significant proportion of older women who basically have been living okay in small towns for years until their husbands died. They don't really have any kind of death benefit programs or insurance. Their husbands probably worked at minimum wage. They are finally left to themselves on a $191-a-month Social Security check. These are the people on the edge."[37]

Mary Farwell, director of a church-based rural support group in eastern Iowa, got her first view of life on the edge as an emergency-room nurse in 1984. "There was one older woman who came into the hospital with a leg almost lost to gangrene," Farwell recalls. "How could anyone let an injury go that far? I wondered. What was she thinking of? I found out later that she didn't have any insurance—she couldn't afford it. But she was only the first. I saw people die, I saw many people die, because they didn't have the money to pay for a doctor."

For the rural poor, health insurance is a luxury, as far out of reach of their meager incomes as a trip to Hawaii or a new car. First comes food, next comes shelter, then comes heat. In a growing number of homes, the list of affordable expenses ends there. Preventative medical attention, such as annual check-ups, pap smears, and the like is simply nonexistent.

The problem is widespread throughout the Heartland. More than half of the 1,800 rural doctors in Iowa, Kansas, Missouri and Nebraska who responded to a 1986 survey reported that patients were waiting until later stages of illnesses before seeking medical attention.[38] The problems created by inadequate health insurance are compounded by the fact that rural people in general have greater need of medical services, and the rural poor—the least likely to have insurance—are the most needy.

"The first thing rural people give up when things get tight is health insurance," says Barb Grabner, a staff worker with the Iowa-based Farm Unity Congress. "I see many people who need treatment, but who can't afford it. That's a crime in this society. There's no need to let this happen."

Roseann Johnson, now of Des Moines, learned what it means to lack insurance when she was living in rural Kansas. Divorced and raising her five-year-old daughter Heather by herself, Johnson felt she couldn't afford health insurance. One night, when Heather had trouble breathing, Johnson called for an ambulance. The ambulance attendants refused to take Heather to the hospital when they found out that Johnson lacked insurance. An hour later, the child went into convulsions and died. It was learned at an autopsy that Heather had died of meningitis, a disease that is generally treatable. Such horror stories have become more common in Iowa, where a third of all state residents lack sufficient health insurance, and where a third of those without any coverage at all are children.[39]

We are all familiar with scenes from television newscasts of the well-heeled stepping over the bodies of the homeless poor on the sidewalks of America's largest cities. These images of tremendous wealth existing side-by-side with abject poverty make for compelling TV: they are striking and evocative, and they neatly sum up harsh ironies in American society. But such images are, perhaps, a bit *too* neat. They oversimplify what is in reality a complex picture of poverty in this country, obscuring the fact that the line dividing the affluent from the poor in America is increasingly being drawn geographically, with the rural Heartland falling farther and farther behind urban coastal areas, prompting some economists to talk about the development of a "bi-coastal" economy.[40]

Even within the Heartland itself, the economic gap between urban and rural areas has widened to record levels, and it is increasing. What may prove to be a permanent underclass is silently taking root in rural ghettos. Given the increasingly interdependent nature of American society, the effects of rural decline will soon be felt in even prosperous urban communities. Our cities, for example, simply cannot absorb new waves of refugees fleeing rural ghettos, as they have so often done in the past. The truth is that, as sociologists Don Dillman and Daryl Hobbs have pointed out, "Neither the urban nor rural portions of our society can flourish for long while the other languishes behind."[41]

Perhaps the most frightening aspect of this economic schism is that it has developed in an era of general economic growth. The rising economic tide that was supposed to lift all the boats in the harbor floated some while swamping others. And there are other gaps springing up in rural America: people who used to cooperate now see each other as competitors. One farmer's hard times are a neighboring farmer's opportunity to expand by buying up land at fire-sale prices. It is very difficult to help your neighbor out when the key to your success—and at times, your very survival—appears to be his or her destruction. This emotional distance, and active antipathy, between the successful and the unsuccessful is found increasingly within small towns. According to rural sociologist Cornelia Butler Flora, these schisms are the result of severe economic distress pounding bedrock rural values.

"Rural people have always believed that if you work hard, you will succeed," says Flora. "But for many that just isn't the case. In rural areas, more so than in urban ones, the individual is viewed as being responsible for his or her poverty. That splits the community apart."

Johnny Baker, the owner of a service station in a small eastern Iowa town, is a true believer in the Midwestern work ethic. We meet at the local Elk's Club, where we are seated next to each other at a dinner party. Johnny is a friendly, open-faced man with a gift for easy conversation. When he finds that I am interested in the effects of the farm crisis on rural communities, his eyes narrow a bit but he says nothing, and we move to a different topic. Throughout the meal I catch him glancing over at me, his eyes darting away when I look in his direction. As dessert is being served, I suddenly feel his hand on my arm. With what must be his third highball sitting before him, he says forcefully, if a bit unsteadily, the complaint that has obviously been preying on him all night: "I can't feel too sorry for a bunch of farmers who have been getting handouts from the government for years when I never got a dime from Uncle Sam."

A month later, on the opposite side of the state, I am sitting with farmer Joyce Ebbers in her kitchen, listening as she tells the now-familiar story of how her family struggled—unsuccessfully—to hold onto their farm. Fast expansion pushed by farm lenders and the United States Department of Agriculture, skyrocketing interest rates, declining profits, desperate times: a dream shattered.

The conversation turns to the reaction of the nearby townspeople. "Oh, it's a disgusting town," she says, her eyes flashing. "At the beginning of the farm crisis reporters would come out and interview us about what was going on. One lady from town came up to me while I was shopping one day and said, 'If you want to talk about your problems on the news that's up to you. But do you have to say you're from here?'

"Well, I don't shop in town anymore," she continues. "I go to Des Moines. And that lady who said those things to me? She owned a furniture store in town and the roof just caved in on it a month ago. Destroyed the whole store. Serves her right."

There has always been a gap separating farm people from those in town. The residents of rural communities often see themselves

as being slightly superior to, a bit more advanced than, their
neighbors on the farm. Even in the smallest rural schools, farm
kids are made to feel different. For their part, many farmers
harbor a gut-level belief that farming is the only "real work"—that
the shopkeepers, doctors, lawyers, and schoolteachers in the
nearby community live off of the farmer's honest sweat. But the
farm crisis of the 1980s has greatly increased the traditional gulf
between town and farm, as well as that between the successful and
the unsuccessful.

When asked how his rural parishioners are faring, Father Robert Thomas, a Catholic priest in Iowa, just sighs and reaches into
his desk drawer. He pulls out a stack of slips of paper the size and
shape of note cards, held together with a rubber band, and drops
the thick pile onto his desk.

"These are requests . . . no, *pleas* for help from our rural parishioners," he says. Father Thomas, a small man with thinning
gray hair and a deeply lined pleasant face, picks up the pile of
cards, removes the rubber band, and starts reading from them at
random: "Dental bill: $100. Doctor bill: $45. Glasses: $100. Phar-
macy: $50. There are over 100 of these slips here. The need for
medical care is just tremendous, and it is going unanswered in too
many cases. Imagine that: a lack of medical care, here, in America, the most 'advanced' country in the world. And people are
actually dying because of it! The destruction of these people's
lives is a sin—pure and simple."

Father Thomas removes his glasses and sets them carefully on
his desk beside the large stack of slips, each one containing a
name, a phone number, a dollar amount, and the reason for the
request. He must choose which requests are the most deserving
and so merit help. Those who sent the other requests, the not
quite as desperate, will have to muddle through somehow.

Father Thomas knows rural people well. He was born in Iowa,
the son of a small-town banker. When his father's bank closed
during the Depression, the family moved to a farm. Father Thomas has seen a lot of suffering over the past several years as the
sinkhole of poverty opened up beneath many of his parishioners.
He has sat up nights with farmers who had just lost the spread
their grandparents had carved out of prairie land more than a
hundred years before. He has taken boxes of food and children's
clothing to countless homes in small towns and out in the coun-

tryside. He tells about the time a bankrupt farmer went hat in hand to his banker seeking permission to withdraw enough money to buy his teenaged daughter a prom dress. When the girl found out what her once-proud father had been forced to do, she drove the family car off a nearby bridge, almost killing herself.

But of all the suffering Father Thomas has witnessed, it is the loss of compassion for those in trouble, what he calls "soul erosion," that bothers him most. "It's a corrosive attitude," he says dolefully. "It's frightening to behold, and I'm seeing it more all the time."

After leaving Father Thomas's office, I learn just how difficult it is for those who must watch their communities fall into hard times. In the elevator, a secretary asks what I think of Father Thomas. When I say that he is quite a man, she nods emphatic agreement and adds that "this whole thing" is just killing him. "What is?" I ask.

"This farm thing," the secretary says. "Didn't you know about his stroke?"

I shake my head.

"Oh, yes," she says. "We worry about him. And did he tell you about what happened to his head?"

I had noticed the large Band-Aid on his forehead but hadn't asked Father Thomas about it.

The secretary explains: "Father was sleeping and he dreamed that somebody walked up to him and said that the farm crisis was the farmers' own fault. Father couldn't stand it, and he tried to punch the man. He woke up on the floor with a gash in his forehead. He had actually swung at the man in his sleep and had fallen out of bed. He hit his head on his bedside table. That should tell you something about how this whole thing is affecting him."

Driving home, I think about Father Thomas, who is the picture of priestly mildness, being so troubled by the rural decline that he actually jumped from his bed trying to silence the nightmare voice of indifference. If that is the effect on someone who merely witnesses the devastation second-hand, what, I wonder, is it like for those who find themselves and their families tumbling down into a dark hole where only the moment before existed a life bright with the promise of the American dream?

5

The Dying of the Light

There's so much pain around here that it's hard to rise above it.

KATHY LEHRMAN,
Mechanicsville Newspaper Publisher

Hills, Iowa

Long before the bitter-cold morning of December 9, 1985, Dale Burr must have realized that he was in over his head. The first blow came early, back in 1972, when Burr's father, Vernon, died.

According to Dale Burr's brother-in-law, Keith Forbes, it was the elder Burr who made the principle decisions for the father-and-son farming operation. After Vernon Burr died, "Dale started slipping," said Forbes. "His crop was always in late and out late. He worked hard all the time, but he was just spinning his wheels. He needed management."

On the morning of December 9, when most of his neighbors had harvested their entire crops, only half of Burr's 150 acres of soybeans and three-quarters of his 250 acres of corn had been harvested. The rest of the crops stood in the snow-covered fields.

Despite his problems, the 63-year-old Burr was known to friends and family as an easygoing, happy man who loved farming.

"He always looked on the bright side," said Forbes. "Always had something good to say about everybody."

In 1982, in what proved a foolhardy effort to stay afloat,

Burr bought 160 acres of additional farmland, paying an average of $3,500 an acre. Land values dropped almost immediately. As times grew harder, Burr began borrowing heavily from his mother to make land payments. Then in early November 1985, Burr became embroiled in a dispute with his bank and a government agency over a $23,085 loan he had taken out.

On the week before December 9, Burr and his wife Emily drove over to visit her sister, Ruth Forbes, and her husband, Keith. That in itself was a rare event.

"They just never, never, never did that," said Keith Forbes. But there was another surprise. Burr, a quiet man who kept his problems to himself, unloaded his troubles on the Forbeses, telling them about the massive debts owed to the bank and to Burr's mother. He owed nearly $800,000 to the Hills Bank alone. He had no idea how he could repay the money. Now the bank was threatening to foreclose on him. "He told me he didn't have enough money to buy groceries," Forbes recalled. "His pride was broken."

Mary Farwell, director of Farmer's Outreach, a church-based rural support service in eastern Iowa, vividly remembers the meeting held in the basement of the Hills Catholic Church during the first week of December 1985. The men, dressed in insulated coveralls and wearing seed-corn hats, stood around talking quietly in groups of two or three before the meeting, edgy and grim. The women sat on the metal folding chairs provided for the meeting and said little, their hands folded on their laps. The support group for farm families to share their troubles with neighbors was one of only a handful operating at that time in the Midwest. The meeting went as well as could be expected with farmers, who as a group still hold fast to the rural sentiment expressed in Martha Gellhorn's Depression-era novel *The Troubles I've Seen*: "Hard times were something you managed by yourself."

After the meeting, a woman walked up to Farwell and her assistant and without a word laid her head on the assistant's shoulder and began to sob.

"She was so distraught," Farwell recalls. "She said one of her relatives was having a bad time. My assistant took the woman to the side and talked with her for a long time, until she calmed down. We assumed that the local people would step in then and keep track of things. I remember on the way home, my friend and I agonized over what had happened. Should we have said more? Less? Different? A week later—just days after December

9th—we found out that that woman was related to Dale Burr."
Farwell leans forward in her chair, her eyes suddenly open
wide. "She had been talking about him."

The morning of December 9, 1985, dawned cold and gray in
Iowa City, home of the University of Iowa. With a population
of 50,000, Iowa City is as cosmopolitan as towns get in Iowa. It
boasts several upscale restaurants, a handful of excellent book-
stores, the country's most prestigious writers' workshop, and a
center for the performing arts that serves as the summer home
for the Joffrey II ballet company. It has been called, only some-
what humorously, the Athens of the Midwest.

The town of Hills, population 550, is just seven miles from
Iowa City, one of a dozen small farm towns clustered around
the larger community like moons around a planet. But despite
the geographic proximity, Iowa City has more in common with
New York City or San Francisco than it does with Hills. Mention the college town to farmers only a few miles away, and
they'll shake their heads and recommend you steer clear of that
modern-day Sodom. Mention Hills to the people in Iowa City
and you'll get blank looks.

"It's difficult for people in [Iowa City] to appreciate what's
happening outside," an agricultural economist told a reporter
who called on the afternoon of December 9.

In the early morning hours of December 9, just seven miles
across barren fields from the bustling community of science and
art, the sinkhole that had been growing silently and steadily beneath Dale Burr for over a dozen years finally swallowed him
up. He stood in the kitchen of his modest farm house, hastily
jotted down a short note, and then laid it on the counter. "I'm
sorry," the note read. "I can't take the problems anymore."

He took out his old 12-gauge pump-action shotgun and shot
his 65-year-old wife, Emily, once in the chest at close range,
killing her instantly.

He walked outside, climbed into his green and white Chevro-
let pickup and headed into town, just three miles south. Friends
of Burr's at the Hills Grain and Feed recall him driving past
and waving to them at about 11:15 a.m.

Burr parked his truck, walked into the Hills Bank and Trust,
and tried to cash a check on an account that was long over-
drawn. The teller informed Burr that he'd have to talk with a
bank officer. Instead, Burr walked out of the bank, returned to
his truck, put the shotgun inside of his heavy coveralls and
walked back into the bank. He went directly to the office of the
bank's president, John Hughes, 46.

John Hughes was a tall, good-looking man with a reputation for being sympathetic to farmers. Hughes was himself born and raised on a farm, before attending Iowa State University and later graduating from the University of Iowa Law School. The small bank in Hills had grown dramatically during the ten years Hughes had been president.

Hughes was meeting with another bank officer when Burr pulled out his shotgun, opened the door a crack, and shot Hughes in the face, killing him.

Burr closed the door and headed to the rear exit of the bank, passing two bank employees on the way. He aimed his gun at them and pulled the trigger. But the gun didn't fire. Burr had accidentally ejected a live shell instead of the spent one after shooting Hughes.

Burr left the bank, returned to his truck, and drove east out of town, smiling and waving to a friend he passed on the street.

He drove to a farm belonging to his neighbors, Richard and Marilyn Goody, about two miles from town. Stopping for a minute out on the gravel road that runs by the Goody farm, Burr fired a shot out his window, testing the gun after what seemed like a malfunction at the bank. Then he drove into his neighbors' yard, where he found Richard Goody working near the barn.

The two men had been involved in a land dispute which went back to 1982, a dispute which Burr had lost in a court battle. Just as Burr was approaching his neighbor, Goody's wife Marilyn pulled into the driveway. In the truck with Marilyn Goody was the couple's six-year-old son, Mark. Burr trained his gun on 37-year-old Richard Goody, who had survived a tour of duty in Vietnam with an Army artillery unit, and fired twice. Goody fell back into a snowbank between two hog feeders, dead.

Burr then turned his attention to Goody's wife and son. Marilyn Goody pushed her son down to the seat of the truck as Burr raised his gun to his shoulder and took aim. She jammed the gas pedal to the floor and sped off down the gravel driveway. Burr fired at the fleeing truck, but it was already too far away. A score of small dents made by shotgun pellets on the side of the Goody truck proved how narrow Marilyn and Mark Goody's escape had been.

Ten minutes later—at 11:45—Burr's pickup truck was spotted by Sheriff's Deputy David Henderson on a snow-covered county road five miles south of the Goody farm. Henderson signaled Burr to pull over by flashing his lights and sounding his siren. The farmer did as he was instructed. The two vehicles sat on the side of the road while Henderson waited for backup help from the Iowa Highway Patrol and the county sheriff's department. When the extra units arrived, officers walked carefully up to Burr's truck. They found Burr, slumped forward, dead from a single self-inflicted shotgun blast to the chest.

When the news of the murders hit the Hills Tap bar, one patron was incredulous that Dale Burr could have done such a thing.

"He was a hell of a nice guy," the farmer said, stunned. "He go nuts?"

"He went broke," answered someone in the bar.

Later that afternoon, a man called the Iowa State Bank in nearby Iowa City. "Tell Marty he's next," the anonymous caller said, in an apparent reference to the bank's agricultural loan officer.

A week after the killings, Mary Farwell's assistant, the woman who had tried to comfort Dale Burr's relative, quit the farm support group. "I just can't handle the responsibility anymore," she told Farwell, crying.

Three months after the shootings, James Gordon, a bank officer at the Hills Bank who had been in the building the morning Dale Burr shot John Hughes, died of a heart attack at the age of 38. His family was convinced that the murders in Hills contributed to Gordon's early death.

The murders in Iowa were not the first cases of violence connected to the farm crisis, nor were they to be the last. But for many Americans, both rural and urban, the killings represented a watershed. In the words of Mary Farwell, "We suddenly realized how deadly serious the game was."

Burr's rampage was front-page news in papers across the country. "Death on the Iowa Prairie: 4 New Victims of Economy" was the headline the *New York Times* gave to its article on the murders.[1] *USA Today*'s read, "Iowa town's grief: they understand."[2] And *Time* declared, simply, "He couldn't manage any more."[3]

For a brief and awful moment, the farm crisis was *the* hot story. It was almost impossible to turn on the evening news without being confronted by the image of a farmer crying as the family farm was auctioned off. But before long the cameras turned to other crises, and soon the very concept of a farm crisis disappeared from public view. Despite the lack of media attention, murder-suicides involving farmers and their families became, if not commonplace, at least regular. On New Year's Day 1986—less than one month after the Burr shootings—an Idaho farmer in financial trouble shot and killed his wife, his 16-year-old step-daughter, and himself. In South Dakota a week and one-half later, a small-town banker left a note stating that his job of dealing with troubled farmers "has got pressure on my mind" and then shot to death his wife and two children before killing himself. In August of the same year in Oklahoma, a man who had recently lost his farm to foreclosure shot and killed his wife and two children before setting fire to their home and killing himself.

But suicides (without the added horror of a family massacre) had already become widespread by the time of the Dale Burr killings, and the numbers continued to swell afterwards. In 1987, the number of suicides in Iowa climbed to 398, the highest number since the Depression. (One hundred and ninety-five Iowans shot themselves; 74 used poisonous gas or vapors; 66 hanged or suffocated themselves; 35 took poison; and 28 died by a variety of other methods.[4]) The suicide rate among farmers in Iowa was 46 per 100,000 in 1983. The national rate for all adult men is about 29 per 100,000.

The number of rural suicides is vastly underreported, the deaths often attributed to farm-related accidents by family members, local doctors, and county medical examiners. During the farm crisis, a suspicious number of farmers died in hunting accidents—while in their own barns. One farmer was found hanged in his barn. The official cause of death: heart attack.

Multiple murders and suicides are, of course, only the most extreme, the most visible evidence of what some experts have called a "mental health crisis" that has enveloped the rural Midwest. For all the outer changes that have taken place in rural America over the past decade, it is the change occurring inside the hearts and minds of rural people that is the least recognized and, perhaps, most important. The constant downward ratcheting of expectations, the grinding, daily battle against largely unknown but seemingly invincible enemies, the dissolution of families, communities, and dreams are taking a toll on rural people that will last for decades.

"When a fellow in a steel mill loses his job, he has basically lost his paycheck," a professor of preventive medicine at the University of Iowa recently said. "When an Iowa farmer loses his farm, he's lost the guts of his life."[5]

The first half of the doctor's assertion is an understatement. A job—any job—is much more than a paycheck; the experience of downward mobility is strikingly similar whether it is endured by farmers or by others who have lost their jobs. As anthropologist Katherine Newman points out, "In addition to money, a job confers prestige and a sense of purpose. . . . [The downwardly mobile] must therefore contend not only with financial hardship but with the psychological, social and practical consequences of 'falling from grace,' of losing their 'proper place in the world.' "[6] One study found the suicide rate among a group of displaced workers—not farmers—was 30 times higher than the expected number due to the psychic traumas endured by laid-off workers and their families.[7] Clearly, it is wrong to minimize the suffering of white- and blue-collar workers. Still, there are some important differences between losing a job and losing a farm.

According to Bonnie Williams, a professor of social work at the University of Iowa, who grew up on a farm, "The farm is part of the family. My great aunt and uncle had our farm originally; my father has lived there since the day he was born. If he lost that farm, he'd feel as if he were failing the whole family. It would say that not only was his life meaningless, but that the lives of the preceding generation were meaningless too."

To fail several generations of relatives (both backwards and forwards into those unborn descendants who will now not be able to farm), to see yourself as the one weak link in a strong chain that spans more than a century, is a terrible, and for some, an unbearable burden.

"It took a good year of talking for me to convince my husband to file for bankruptcy," said one farmer. "He thought he was a failure. He was afraid that people would talk. Several other people from around here who filed bankruptcy just picked up and left in the middle of the night without telling anybody."

There are other important differences between losing a farm and losing a job. When a feeling of group solidarity exists—as in a union—workers who have been laid off experience fewer mental and emotional problems. In the case of the striking air controllers fired by President Reagan in 1981, their union affiliation with the

Professional Air Traffic Controllers Organization (PATCO) kept them from the kind of mental anguish that results in suicide and murder.[8] The union provided a sense of community that sheltered the former air controllers from self-doubt and self-blame. Their loss of jobs and status was seen as the result of a larger battle between contending social, economic, and political forces. In short, the PATCO workers didn't take their injury personally.

But the same cannot be said of farmers. There are no significant farmers' unions providing a similar group identity to those who have lost the family farm. Quite the opposite: the myth of the fiercely independent yeoman farmer, long held as the ideal, makes farmers into competitors, not cooperators. When a farmer goes down, he or she goes down alone. The same macho myth of independence has also made it almost impossible for farmers—especially men—to admit to having problems and asking for help.

"They just don't want to talk about their problems openly," says Joanne Dvorak, a mental health professional in Cedar Rapids, Iowa, who has worked with farmers for several years. "They are strong, proud people. They want to put their best foot forward, and when there is no best foot to put forward, they stay home."

Patty and Larry Thomann fit Dvorak's description perfectly. Until 1985 the young couple owned a medium-sized farm, just a few miles away from Dale Burr. When they were forced to sell out, the move surprised even their families.

"We were artists at not letting anyone know that we had problems," says Larry.

"You not only don't let your friends and family know what you're feeling," Patty adds, "but you deny it to yourself. You have to keep telling yourself that it's going to be all right, even when you know it's not, because otherwise you're sitting on a time bomb. You have to deny it just to survive."

Not all of these human time bombs go off in a single violent explosion, as in the case of Dale Burr. But the corrosive effects of a slow and relentless psychic burn can be just as destructive. If you can get past the Heartland facade of tranquility, if you can see through the mask of openness that Midwesterners have worn for so long that they themselves confuse it with the real thing, it is abundantly clear that the soul of rural America is suffering.

Look first to the children, the most vulnerable and the most expressive part of a family. The image most Americans have of country life in which children grow up happily and healthily is sharply at odds with today's rural reality. Over the past decade there has been a dramatic rise in many indicators of social distress among rural youths, factors such as teen pregnancies, alcoholism and drug use, runaways and suicide. Social workers tell stories about farm children who hide their favorite toys for fear they will be taken away by the bank—just as their parents' tractors and cars were repossessed.

"One ten-year-old said he couldn't sleep at night because he was worried that a truck would come and take his parents away," says Joan Blundall, a counselor in northwest Iowa. "He had, after all, watched strangers come and take the family's hogs."[9]

A 1986 study of 4,300 rural Minnesota adolescents found that three out of every hundred attempted suicide in the previous month—a figure 15 times higher than the national average. In a standardized test for depression, 18% of these rural adolescents were found to be moderately or severely depressed. Worse, the average score on the test—the higher the score, the more serious the depression—for these troubled rural youths was greater than that for adolescents hospitalized at the University of California at Los Angeles Neuropsychiatric Institute at the same time.[10]

The burden of depression is especially harmful to adolescents, who are going through a period of great change, a time in which they form essential parts of their identity. Children in homes torn by economic hardship often become the targets of their parents' pent-up frustrations. As the rural economy has slumped, the number of cases of child abuse involving rural children has grown at a rate even greater than that involving urban children. Despite a shrinking population, the number of child-abuse cases reported in Iowa went from nearly 15,000 in 1979 to over 25,000 in 1987.[11] (Some of this increase is no doubt due to improved reporting procedures, but experts agree that the increase reflects a significant rise in the number of actual cases of child abuse.)

There has also been a dramatic increase in domestic violence against rural women in just a short time. In Iowa, reported cases of spouse abuse went from 1,620 in 1985 up to more than 4,500 in 1987.[12]

"People seem to think that domestic violence is just an urban problem," says Jennifer Etter McCoy, volunteer coordinator of Domestic Violence Alternatives, which works with battered women in central Iowa. "If anything, the problem is worse in rural areas because it's even more of a hidden problem here."

Because of the rural prohibition against complaining about personal problems (and especially against seeking outside help for them), battered rural women are even less likely than urban women to report incidents of abuse. "You just don't air your dirty laundry in public, if you're a rural woman," says McCoy. "Rural women are very, very isolated."

Even if a battered rural woman wants to get help, she faces problems not encountered by city residents. To whom can she turn for help? Where can she find safety? Shelters for battered women are extremely rare in rural areas. So are counseling centers. Few of the rural clergy are trained to deal with domestic abuse. A battered woman may be afraid to tell a neighbor for fear that the information will get out into the community, causing her to be shunned, or worse: her abuser may punish her for revealing their "home situation." Because of these fears, rural women are often forced to travel great distances for help. Because of the difficulties in making such a trip, many rural women simply try to "tough it out," leaving their lives in jeopardy.

Access to mental health services of all kinds is extremely limited throughout rural America, and as is so often the case in rural areas, those who need these services the most have access to them the least. In 1985, Iowa—the state hardest hit by the unraveling of rural life—had the lowest per capita funding for mental health agencies among the 50 states.[13]

A large part of the problem can be traced back to the Reagan revolution. Since 1981, the federal government has slashed the amount of federal funds available to the states for social service programs. In 1980, the states received $314 million for operating community mental health centers. That figure declined to $204 million in 1982, and then rose to just $238 in 1988 (and these numbers have not been adjusted for inflation).[14]

This policy of federal neglect for social services came at the worst possible time for rural Americans: at the precise moment that the rural sinkhole began to crumble and rural people needed those services the most. Given the one-two punch of reduced funding and increased need, states have had to take drastic actions. According to an official with the Iowa Department of Human Services, "Most social workers are drawing up a priority list of the most dangerous cases of child abuse and neglect. Other cases must be neglected."

The result isn't hard to imagine. "We are seeing a mental health disaster throughout the Midwest," says JoAnn Mermelstein, professor of social work at the University of Missouri in Columbia. "Whole communities are immobilized by fear, apathy and an inability to see any options. It's splitting apart families in a terrible way. There is enormous pessimism about the future out there."

In a survey of community leaders in one five-county rural area, Mermelstein found that 93% of the leaders believed that everyone, or almost everyone, in their community was suffering. Three out of four of these same leaders believed that their community was dead or dying.[15]

This loss of hope is one of the keys to understanding the change that is taking place in rural America. It represents the loss of faith in the American system's ability to ensure a meaningful and secure position to those willing to work hard. That shining light of the American dream has gone out for these rural people.

Says rural advocate Mary Farwell, "These people—the most patriotic, idealistic, Heartland Americans—have lost faith not only in themselves but in their country. Because when they were hurting they were ignored. That does something to a person, something that's frightening to see."

Jaymee Glenn-Burns, a pastor at the United Methodist Church in Mechanicsville, Iowa, does her best to give the members of her congregation hope. The sermon she delivers on a crisp October morning relates the Old Testament story of Ruth and Naomi. But the story of the deep friendship between a mother and her daughter-in-law is just a vehicle to ride into the theme of hope. She uses the word "hope" four times in the opening minute alone.

She ends her sermon by telling her congregation, comprised of farmers and ex-farmers, small-business owners and former owners of small businesses:

> Friendship, faith, and promise. What do these do for us today? Like Ruth and Naomi we are victims of things we can't control, that bring hardship, pain, and sorrow into our lives. Many of the causes of our suffering are things we have no control over—another person's feelings or actions; arthritis, cancer, depression; the weather; the economy or the values of a changing society, or a community that can never be what it once was.
>
> What's left for us? What do we do when there's nothing we

can do? We trust God. We love our friends and encourage one another. We hope, even in the midst of despair, and cling to the promise that God is even now in the process of redeeming our suffering, restoring our broken lives, and saving the world.

Glenn-Burns is young and exuberant. Her vitality is like a flame from which others can draw strength. But there are days, she admits, when the pain of the community is too much for her.

"There is so much anger and sadness," she explains, sitting in her church office. "Sometimes it's overwhelming, and I go into my room, close the door, and scream." She sits quietly for a minute, weighing her words before she speaks again. "I think we're groping," she says at last. "We don't know exactly where we're going, and somehow we're struggling to keep something of our community intact as we go. We're just letting things happen, and hoping that it all turns out okay. We're hoping for the best."

6

The Growth of Hate Groups

After God and country and neighbors desert you, what's left?

PRIEST IN RURAL IOWA

Colony Village Restaurant, off Interstate 80, Iowa

To reach the basement meeting room at the Colony Village Restaurant in eastern Iowa you must first run the Norman Rockwell gauntlet—past shelves crowded with porcelain figurines of solicitous country doctors and red-haired boys toting fishing poles, refrigerator magnets in the shape of pigs, racks heavy with a cornucopia of apple butter, plum butter, pumpkin butter, honey-glazed hams, shiny black licorice ropes and red licorice pipes, jars of cole-slaw dressing, and a dozen different varieties of sauerkraut.

Few locals buy these self-consciously Midwestern items. The souvenirs that spill out of the large gift shop and cover the walls from the entrance of the building to the coffee-shop door are there to tempt travelers, the families who pull in off nearby Interstate 80, hot and tired from retracing the pioneer trail across the interminable prairie and in dire need of food, drink, bathrooms, and—if the owners of the Colony Village have figured it right—a souvenir or two typical of Iowa.

Far more characteristic of Heartland America is the gathering that is about to get underway on this particular Sunday

morning in the large basement meeting room just below the
gift shop. All of the nearly 100 people here are white; most are
middle-aged or slightly older. They have come in groups of two
and three, passed the knickknacks and come down the stairs,
entering the room through a doorway above which hangs a
wooden sign inscribed with a German prayer:

> O selig haus,
> Wo man dich aufgenommen,
> Du wahrer Seelenfreund,
> Herr Jesu Christ
>
> (Oh, blessed house,
> Where you have been accepted,
> You true friend of the soul,
> Lord Jesus Christ)

The men mill around, talking, joking, slapping each other on
the back, complaining about commodity prices, the weather, or
their wives. The women sit in groups on metal folding chairs,
chatting easily, pocketbooks on their laps. The men wear sports
coats or are in shirtsleeves; few wear ties. The women are
dressed in rather drab floral-print dresses, as if they had called
each other up the night before to agree on what to wear.

At a few minutes after ten a large bald-headed man with a
florid face glances at his watch and exclaims "Oh!" He hurries
to the front of the room and holds up his hands for silence.
When the buzz of a hundred conversations continues, he announces in the booming voice of a minister or a football coach,
"Now, let's get this meeting started, folks. Come on, now. We
have a lot to cover today. Everybody sit down." After a pause
he adds, "Bob, that means you too," a comment that receives
scattered laughs as a red-faced Bob falls into a chair.

The scene is straight out of a Norman Rockwell painting, but
the meeting getting underway is in some ways closer to the vision of another Rockwell—George Lincoln Rockwell, the self-
proclaimed Führer of the American Nazi Party.

This is the monthly gathering of the Iowa Society for Edu-
cated Citizens—or ISEC—one of several dozen groups to have
sprung up throughout the Midwest in the last decade that are
dedicated to the proposition that the United States is going to
Hell in a handbasket due to the machinations of an international cabal of Jewish bankers and their liberal atheistic friends
in Washington. ISEC was started by Harold Francisco (the bald
man with the loud voice) back in the mid-1970s under the
name "The Patriots of the Constitution."

The wide variety of pamphlets, books, and reprints of articles stacked in neat piles on a table at the back of the room is indicative of the wide sweep of issues taken up by ISEC as it seeks to educate Iowans. The literature decries race mixing, gun registration, the liberal (i.e., Jewish) media, the IRS, homosexuality, the Council on Foreign Relations, and driver's licenses—the last because by accepting them citizens are, in effect, legitimizing what the self-proclaimed patriots consider illegitimate authority. But the target of choice for ISEC members, judged by the number of booklets devoted to unmasking the institution, is the Federal Reserve Bank: root of farmers' problems and the front organization for the international Jewish bankers.

Francisco himself is savvy enough of public relations to avoid making any blatantly anti-Semitic statements during ISEC meetings, referring instead to the "international bankers" and advising group members to do likewise.

"It doesn't look good for our group," Francisco once cautioned group members after one of them had told a TV reporter—on camera—that he thought Jews should be killed. "That's bad," Francisco warned members. "We don't need to make statements like that. Even if you believe it, don't say it."

Ed Murphy, the offending member, had made his controversial statement during a special segment of a local Iowa TV newsmagazine which was devoted to the growth of far-right groups such as ISEC. In a chilling exchange, Murphy, a grizzled but mild-looking man of around 40, wearing an embroidered cowboy-style shirt, calmly states that "Jesus says Jews are his enemies," adding that he would do anything to protect his race and religion. The reporter asks if such protection would include killing Jews. Murphy hesitates only a second before replying, "Well, yeah. Get rid of them. [Jesus] says get rid of them. I say get rid of them."

On this particular Sunday morning Ed Murphy sits toward the back of the room cleaning his fingernails as the meeting gets underway. When everyone is seated, Francisco announces they are lucky to have with them a distinguished speaker: Deloris Kirk, widow of Art Kirk, a Nebraska farmer who was killed—"executed," says Francisco—by a state highway patrol SWAT team two years before.

"Dee" Kirk is a slight, middle-aged woman with brown permed hair and large glasses who has made a crusade out of "setting the record straight about what they did to Art," traveling to far-right meetings throughout the Midwest. The audience is rapt—many of them holding tape recorders—as she calmly begins recounting her version of the events of October 23, 1984, the afternoon three Hall County sheriff's deputies showed up on the Kirk farm just outside of Cairo, Nebraska, to serve legal papers.

Like many other farmers in the Midwest, the Kirks were heavily in debt—they owed more than $300,000 to the Norwest Bank in nearby Grand Island—and had fallen far behind in their loan payments. And as was true for many other farmers, the prospects of ever paying back the loan grew more remote daily.

In May of 1984, a desperate Art Kirk stumbled upon what he believed to be the answer to his problems. At a meeting of a group called Nebraskans for Constitutional Government, Art Kirk sat spellbound as a guest speaker outlined a package of complicated legal maneuvers that could allegedly void the family's loans.

"Every loan made in the United States since 1974 is illegal," the speaker lectured to the crowd. If indebted farmers followed the man's guidance, staying away from lawyers and instead representing themselves in court, everything would be all right. For Art Kirk, who was faced with the possibility of losing the farm once owned by his father, the need to believe that deliverance was at hand overwhelmed any more rational thoughts he may have had.

The speaker at the fateful meeting was Roderick "Rick" Elliot, the founder and head of a group called the National Agricultural Press Association (NAPA) and editor of a tabloid newspaper, the *Primrose and Cattlemen's Gazette*, based in Brighton, Colorado.

Elliot, a bland-looking man in his late fifties, traveled throughout the Midwest selling NAPA memberships at $30 a shot, touting a low-interest-loan scheme, and telling desperate farmers like Kirk who had already borrowed heavily from "the wrong places" that they could simply dissolve their bank debts by filing a flood of legal documents constructed by Elliot and NAPA consultants.[1]

At first glance Elliot's paper, the *Gazette*, resembled hundreds of other small-town weeklies. Agricultural news competed with sports stories in the front pages; then there were recipes for "Backyard Coney Hotdogs" or "County Fair Coolers" and advertisements for farm supplies, pickup trucks, and the Marine Reserves. What set the *Gazette* apart from most other papers was the inclusion of articles warning its readers of a Jewish plot to rule the world and ads for neo-Nazi organizations such as the Aryan Nations and the National Alliance. One series of articles (actually a reprint of articles published under Henry Ford's name in his newspaper, the *Dearborn Independent,* in the 1920s) was introduced as "a crash course on Jewish-Communism, [which] identifies the real problems in this country—the 'International Jew'—and his Satanic plans to rule the world."[2]

In another issue, Elliot reprinted sections from the anti-Semitic classic the Protocols of the Elders of Zion, noting that "The Protocols contemplate a Gentile world ruled by the Jews—the Jews as masters, the Gentiles as hewers of wood and drawers of water, a policy which every Old Testament reader knows to be typically Jewish and the source of divine judgement upon Israel time and again."[3]

This combination of practical information and hate-mongering apparently struck a chord with farmers. In its heyday in the early 1980s, the paper claimed a readership of 40,000—although that figure was probably inflated to attract advertising.

The Kirks became ardent followers of Elliot. When NAPA opened a chapter in nearby Grand Island, they both joined immediately, and Dee Kirk went to work in the chapter office. But the bogus "legal" scheme developed by Elliot quickly unraveled in court, leaving the Kirks worse off than before. Counting on Elliot's maneuvers to wipe out his obligations, Art Kirk had sold $100,000 worth of corn and cattle, assets he had posted as security for his massive bank loan. At 1:15 p.m. on October 23, three deputy sheriffs showed up with legal papers notifying Kirk that his bank was now claiming the remaining security he had posted—including farm equipment and all unsold crops. Kirk was furious. He was also prepared. Kirk, the deputies testified later, carried a .41 magnum handgun.

According to an investigation of the affair made for Nebraska Governor Robert Kerry by retired District Court Judge Samuel Van Pelt, one of the principle links in a chain of factors leading up to the fatal encounter was "Kirk's misplaced reliance on the 'federal post' of his land."[4]

The concept of the federal post was just one of many "legal" tactics cooked up by far-right hucksters and accepted unquestioningly by many farmers, including Kirk. According to NAPA, government officials could be denied access to an individual's land if signs were erected declaring the property a "federal post." Kirk had a score of these meaningless signs nailed to fenceposts around his farm. When the deputies came on his property, he believed his constitutional rights were being violated.

Kirk later told his wife that he had been unarmed when the

deputies arrived with their guns already drawn, and that he
had grabbed a gun only when he thought his own safety was
threatened. The deputies say that they drew their weapons *after*
Kirk drew his. Regardless of which version is true, no shots
were fired, and the deputies withdrew after telling Kirk that he
should consider himself under arrest.

After the deputies left, Kirk went back to his fieldwork, cutting beans on land he had rented nearby. Thinking things had
cooled down, his wife and some visiting friends left to run some
errands. But far from cooling down, the situation was at that
moment heating up. The deputies returned to Grand Island,
where a warrant was issued for Kirk's arrest. Concerned for his
deputies' safety, Hall County Sheriff Chuck Fairbanks called
over to the Nebraska State Highway Patrol and requested that
they send a SWAT team. A short time later the gravel roads
around the farm were blocked off. Dee Kirk was stopped at one
of the roadblocks and prohibited from returning home. An un-
marked State Patrol Cesna cut lazy circles over the property,
keeping track of Kirk's movements. By 8 p.m., State Patrol
Troop C's SWAT team had taken up positions around the Kirk
farm house.

Such a massive show of strength was an unusual response to
the situation, but Sheriff Fairbanks felt he needed to be unusu-
ally wary. Norwest officials had earlier advised him that Kirk
had been wearing a .45 handgun during a recent conversation
with bank representatives. And there were rumors circulating
that Kirk was a member of the Posse Comitatus, a violent right-
wing group whose most famous member, Gordon Kahl, killed
two U.S. marshals in North Dakota before dying himself in a
shootout with law enforcement officials in Arkansas in 1983.

For the next hour and a half, in telephone conversations with
a local newspaper reporter, his wife, and law enforcement authorities, Kirk alternated between quiet, if tense, conversation
and wild, rambling discourses. He told the reporter that the
Mossad, the Israeli secret police, were behind his troubles.

"You think the NKVD and the Gestapo were ruthless," Kirk
said. "You look up the Mossad and see what they've been involved in."

At another point he yelled over the phone at a State Patrol
negotiator, "Goddamn fuckin' Jews! They destroyed everything
I ever worked for. I've worked my ass off for 49 goddam years
and I've got nothing to show for it. By God, I ain't putting up
with their bullshit now. I'm tired, and I've had it, and I'm not
the only goddam one—I'll tell you that. . . . Farmers fought the

Revolutionary War and we'll fight this son-of-a-bitch. We were hoping to do it in court, but if you make it impossible, then, damn you, we'll take you on your own terms."[5]

Kirk soon grew tired of talking. What good would words do? There were, it seemed, only three kinds of people left in Kirk's world: There were the Jewish bankers and their henchmen, out to rule the world. There were the patriots like Rick Elliot and martyrs like Gordon Kahl who stood in lonely defiance against this plot. And then there were the rest of us—all the slumbering innocents, not yet awakened to the gravity of our situation, lambs that would go off to the slaughter without a struggle. Arthur Kirk had made up his mind not to be remembered as a lamb.

In his report to the governor, Judge Van Pelt concluded that by this point Kirk "had reached such a state of paranoia that it is difficult to hold him responsible for his actions. . . . The events of the preceding nine hours had reduced Kirk to the mentality of a cornered animal."

At around 9:30 p.m., with his face and hands painted camouflage-style, a gas mask strapped to his arm, and carrying an AR-15 semiautomatic rifle converted to fire on automatic, Kirk broke from the back door of his house at a dead run and headed for a nearby windmill. Hidden in a natural depression at the base of the windmill and sheltered by a stack of tree trunks was a large assortment of guns and ammunition.

"Freeze, police!" a trooper yelled at Kirk's running figure. The farmer got off a burst from his rifle before it jammed. Two SWAT team members returned fire.

After several minutes of silence, the SWAT team had deputies shine their headlights at the position from which Kirk had fired. There on the ground, bathed in the glare of the cars' lights, lay the lifeless body of Art Kirk. He had been hit twice, once in the thigh and once in the chest. The coroner's report later concluded that Kirk had lived for several minutes after being hit, before eventually bleeding to death.

The crowd gathered in the restaurant basement sits in stunned silence as Dee Kirk finishes the story of her husband's death; the only sounds in the room are the sibilant workings of the many tape recorders and the faint, bubbly strains of "New York, New York," coming from the Muzak speakers in the coffee shop overhead.

When we think of racist organizations operating in rural America, the group that comes to mind first is the Ku Klux Klan, the

white-supremacist brotherhood which rose from the ashes of the
Confederacy after the Civil War. The image of a KKK member
standing before a burning cross, dressed in white robes with the
infamous pointed hood is, in fact, emblematic of American-born
racial and ethnic intolerance.

But our iconography lags behind the times. After a resurgence
of Klan activity during the late 1970s and early 1980s, the group's
fortunes fell. By 1986 the organization had only half as many
members as in 1982.[6] The average member of a hate group today
is more likely to wear jeans and a workshirt or a suit and tie than
a white sheet. David Duke, who in 1989 was elected to the Louisiana legislature as a member of the Republican Party, typifies the
path the far right has traveled in recent years.

A former Imperial Wizard of the Knights of the Ku Klux Klan,
Duke left that group in 1980 to found the National Association
for the Advancement of White People (NAAWP). With his youth-
ful good looks (Duke was 38 years old when he won a seat in the
Louisiana legislature), tasteful suits, and smooth speaking style,
Duke seems a far cry from the cross-burning Klansmen most
Americans loathe. Duke claims to "reject categorically racial or
religious intolerance and hatred," and says he is not a white su-
premacist, but merely "a white civil rights activist." Before the
Louisiana elections Duke changed his campaign slogan from
"Equal Rights for Whites" to the less objectionable "Equal Rights
for Everyone, Special Privileges for None."

But despite all his protestations, there can be little doubt about
the ugly face of bigotry hiding behind Duke's new-and-improved
mask. His organization, the NAAWP, sells literature denying that
the Holocaust took place, information on eugenics, and tapes of
speeches by the late George Lincoln Rockwell, head of the American Nazi Party. When Duke ran for the United States presidency
on the Populist Party ticket in 1988, his campaign manager was
Ralph Forbes, a former KKK member as well as a past member of
the National Socialist Party of America. And appearing with the
candidate at a Duke press conference in Chicago was American
Nazi Party Vice-Chairman Art Jones.

Duke's real achievement—and that of a large segment of the far
right—has been to make racial and ethnic intolerance palatable to
an ever-widening segment of the American public. They have
learned how to cloak the message of extremism in the language of
moderation. They have given fascism a human face.[7]

The spread of far-right groups over the last decade has not been limited to rural areas alone. News accounts of marauding Skinheads, a neo-Nazi youth movement, have grown increasingly common in Seattle, Dallas, Las Vegas, Chicago—in fact, all across the country. But the social and economic unraveling of rural communities—especially in the Midwest—has provided far-right groups with new audiences for their messages of hate. Some of these groups have enjoyed considerable success in their rural campaign. There were an estimated 2,000 to 5,000 "hard core" far-right activists operating in the Midwest in 1985, with an additional seven to ten sympathizers for each activist.[8]

To understand the growth of far-right groups in rural America and to evaluate the threat they pose, it is necessary to divide the movement into two separate but interconnected camps. White-supremacist groups such as the Aryan Nations and its ultraviolent offshoot the Order (which carried out a string of bank robberies and murdered Denver radio talk show host Alan Berg in 1984) exemplify the most extreme tendencies of this admittedly narrow spectrum of the far right.[9] They are differentiated from organizations such as the ISEC, Rick Elliot's National Agricultural Press Association, and others by their propensity for violence and by their rejection of mass organizing as a tactic. Paramilitary neo-Nazi groups such as the Order depend on the invisibility their small size affords to carry out bank robberies and assassinations.

The raison d'être of groups like the ISEC, on the other hand, is specifically to spread the gospel of their ideas to a wider circle of believers. In pursuit of this goal, they, like David Duke, have had to clean up their image, dropping the most offensive hate-filled rhetoric and using code words—such as "international Zionist bankers" instead of "Jews."

Although the scores of far-right groups that honeycomb rural areas are only loosely knit together, most of them share a Byzantine philosophy that combines "constitutional fundamentalism" with a religious movement based on racist thought called Christian Identity. Art Kirk's belief in the notion of a "federal post" was just one example of the hopelessly convoluted political vision of these rural extremists. Constitutional fundamentalists consider both federal income tax and the Federal Reserve System unconstitutional, the imposition of excessive governmental power over free Americans. Many—especially members of the Posse Comitatus—refuse to recognize any government authority higher than county sheriff ("posse comitatus" is Latin for "power of the county"). Adherents believe that they alone are true American patriots, following the original intent of the framers of the Constitution, who, the theory purports, never really advocated democracy as a basis for government. Democracy inevitably becomes "mobocracy," says the far right. What the Founding Fathers envisioned was a "Christian Republic" in which only true patriots (white, Protestant, heterosexual, property-owning men) would rule. It is to the fulfillment of that dream, and to the repression of all those who do not share their vision, that constitutional fundamentalists are dedicated.

The philosophy of the Christian Identity movement—often called simply Identity—is interwoven with the constitutional concerns of the far right. Identity is based on the premise that white Anglo-Saxons, not Jews, are the true descendants of the lost tribes of Israel. According to Identity proponents, Jews, far from being the chosen people, are really the children of Satan. In the world according to Identity, African-Americans do not even rate the human (albeit diabolical) status of Jews; people of color are considered to be members of a subhuman or "pre-Adamic" species and are often referred to in the movement as "mud people."

Although Identity seems to have burst upon the scene in the last decade, its focus on race as the foremost theological issue is not new. Identity has its roots in a mid-nineteenth-century movement called British Israelism, which combined the new scientific theory of evolution with already existing Biblical conjecture about the fate of the ten lost tribes of Israel. According to British Israelites, the lost tribes migrated from Assyria through Asia Minor and into northern Europe—in particular into Scandinavia, Germany, and the British Isles. The key to understanding and fulfilling biblical prophecies, according to the British Israelites, was for northern Europeans to reclaim their "identity" as the chosen people and to then keep their race pure by not interbreeding with inferior types such as Jews and African-Americans.

In modern Identity thinking, the United States is the Promised Land, the site of the New Jerusalem which will be established after an apocalyptic race war, or "end-time." Many Identity adherents have stockpiled weapons and food in anticipation of this glorious and bloody day.[10]

The dangers inherent in the combination of constitutional fundamentalism, with its populist appeal, and the Identity movement, with its theological rationalization of racism, are obvious. As Leonard Zeskind, research director of the Center for Democratic Renewal, points out, "Identity . . . plays a dual rule: it provides religious unity for differing racist political groups, and it brings religious people into contact with the racist movement."[11]

Indeed, Identity today provides the lingua franca of the far right, binding together members of the Klan, the Aryan Nations, and smaller unaffiliated groups like the ISEC. Identity ministers spread their racist teachings by working the far-right circuit, speaking at Klan rallies, at neo-Nazi paramilitary encampments, and at the countless basement meetings held across America. They broadcast their theories of racial purity and Jewish conspiracies on many of the country's low-wattage AM radio stations— and increasingly on more mainstream stations as well. (For example, Pete Peters, an Identity minister from Colorado whose services were attended by members of the Order, can be heard Sunday mornings on 21 AM and FM stations in 18 states.)[12]

And like their fundamentalist counterparts, Identity ministers make use of modern technology, videotaping sermons and distributing them through catalog sales as well as passing them hand to hand through the far-right grapevine. Many of these videotapes are amateurishly produced, with sound and lighting quality comparable to the cheapest porno films. But others are as slick as anything on network television. There are, as yet, no Identity tapes that use the state-of-the-art attention-grabbing techniques refined by MTV, but that day may not be far off.

Two of the far right's more determined—and in many ways, most successful—forays into the Heartland political arena have been the election campaigns run by followers of the well-known right-wing ideologue Lyndon LaRouche and those coordinated by the new Populist Party, formed in 1984 by the less familiar extremist Willis Carto.

LaRouche is known to millions of Americans for his quadrennial presidential-campaign television speeches in which he very rationally outlines wild conspiracy theories (Queen Elizabeth and Henry Kissinger run the world drug trade) followed by even wilder high-tech solutions (building thousands of nuclear power plants, a giant pipeline funneling water from the Yukon and McKenzie Rivers down to the southwestern United States). But while LaRouche's ideas are fantastic, more importantly, they are also fascistic, grounded in a culture of authoritarian control and the veneration of police-state violence. According to journalist Dennis King, the country's leading expert on LaRouche, the very outrageousness of the man's spiel deflects attention from the real totalitarian threat posed by LaRouche.

Chip Berlet, another veteran LaRouche watcher, agrees with this assessment. He says, "I talked with dozens of reporters. I'd send them LaRouche's writings. Then I'd lead them step by step through it on the phone, to show them it was classic fascism. . . . They'd say, 'That's nice,' then turn to their word processors and crank out some quip about Queen Elizabeth."[13]

Lyndon LaRouche started out on the political left during the 1960s, as a Marxist intellectual on the fringes of Students for a Democratic Society. His journey to the far right began when he advocated adopting the tactics of the fascists in order to combat anyone who got in his way, left or right. In 1973, for example, LaRouche sent club-swinging supporters after members of several left-wing groups in a bloody campaign called Operation Mop Up. Soon after, LaRouche dropped the pose of Marxism and became an unabashed fascist, calling for a militarized society led by a supreme dictator. (According to LaRouche, the individual best prepared to don the mantle of dictator is, not surprisingly, LaRouche himself.) As journalist King points out, LaRouche's fascistic mind set is perhaps most evident in his plan for fighting AIDS, a collection of draconian measures including the imprisonment of anyone who may have been exposed to the AIDS virus—especially gays, prostitutes, and intravenous drug users—and the possible execution of those responsible for spreading the disease. LaRouche's plan bears a startling resemblance to Adolf Hitler's campaign against syphilis, outlined in *Mein Kampf*. And like Hitler's, LaRouche's concern for public health has been in reality simply a means to capitalize on existing public fears to further his own political ends. "*New Solidarity* [LaRouche's newspaper] . . . suggests that AIDS might become the springboard for a nationalist revolution," writes King.[14]

The social upheaval caused by farm problems presented LaRouche with another potential springboard. LaRouche has made a special effort to rally farmers to his cause, beginning with a 1978 campaign to win over members of the American Agriculture Movement (AAM)—the group which had organized the "Great Tractorcade" on Washington, D.C. The AAM presented LaRouche with a perfect opportunity: here was a group of angry people who had already committed themselves to political activism. Under the aegis of the AAM, thousands of farmers riding tractors from across the country had descended on Washington to protest U.S. farm policy. They camped out on the Mall while lobbying Congress for a slew of agricultural reforms. They also caused massive traffic tie-ups throughout the city, clashed with the police, and even set an old tractor ablaze on Independence Avenue. All that LaRouche needed to do was align himself with the farmers' interests and, once inside the organization, steer the AAM in his own direction.

Although LaRouche failed in his effort to take over the farm group—the leadership of the national organization eventually realized that beneath the sympathetic words lay a totalitarian agenda and repudiated LaRouche and his political organization, the National Caucus of Labor Committees (NCLC)—LaRouche did manage to recruit from the leadership of several state AAM chapters. His campaign to portray himself as a "friend of the farmer" (LaRouche chose a Mississippi farmer, Billy Davis, as his 1984 vice-presidential running mate) also earned him the loyalty of several other farm activists, including a former Missouri state treasurer of the National Farmers Organization and a former national board member of the National Organization for Raw Materials.

LaRouche's courtship of farmers is part of a well-thought-out political plan of attack. "The fight we're aiming at is to build an urban-rural alliance," LaRouche lieutenant Peter Bowen explained in a telephone interview. LaRouche has for years dreamed of building a potent mass organization composed of the disenchanted residents of inner-city areas and people of small-town America. To further the urban side of the equation, he has made a significant effort to reach out to the African-American community, visiting black college campuses regularly, forming a National Anti-Drug Coalition, and even sponsoring a Martin Luther King, Jr., birthday march in Washington, D.C.

But for all of LaRouche's ingratiating rhetoric about farmers and African-Americans, he has nothing but scorn for the two groups he claims to represent; they are simply pawns in his grab for power. "Except for a handful of farmers," LaRouche told one audience, "farmers in this country are a bunch of idiots." LaRouche's literature has always been peppered with racist slurs—stating, for example, that black culture is "bestial."[15]

The group's most spectacular electoral victory came on March 18, 1986, when LaRouchian candidates Janice Hart and Mark Fairchild won the Illinois Democratic Party's state primary elections for the offices of lieutenant governor and secretary of state, respectively. LaRouche's wooing of farmers had apparently paid off; a study of county-by-county election returns found a strong positive correlation between support for Hart and a high percentage of family farms in downstate Illinois. Although the pair were defeated in the general election, Hart did manage to garner almost 500,000 votes—certainly a significant showing.

LaRouche also ran a total of 16 candidates in Iowa's 1988 Democratic Party primary. Although none of the 14 candidates in contested races won, they averaged 15% of the vote. Ronald Kirk, a LaRouche-supported candidate for the U.S. House, received 30% of the total vote and captured at least 40% of the vote in five of the sixteen counties in the district. LaRouche himself was the largest third-party vote-getter in Iowa in the 1988 general elections, receiving 3,526 votes.[16]

Just six days before his December 1988 conviction in federal court on charges of conspiring to defraud the IRS and others (a conviction that resulted in a 15-year prison sentence), LaRouche was still at it, trying to win over disaffected farmers at his international "Food for Peace" conference in Chicago.

"The enemy we face is a Satanic movement," LaRouche told 600 attendees in the giant conference hall at the O'Hare Ramada Inn on December 11, 1988. LaRouche blamed the USDA for the drought then gripping the Midwest, and called for a series of changes in federal policy, including an immediate halt to farm foreclosures, a massive infusion of low-interest loans to farmers, and the abolition of the EPA so that farmers could have access to all necessary pesticides, including DDT. One of LaRouche's many skills has always been his knowledge of which buttons to press, and he was clearly pressing all the right ones for farmers.

LaRouche placed the blame for farmers' problems on a familiar cast of villains: the Federal Reserve System, the Trilateral Commission, and "international Zionist bankers." That LaRouche should point an accusing finger at the same "conspirators" as do other far-right groups is hardly a coincidence (or, if it needs saying, proof of the existence of such a conspiracy). LaRouche had been developing ties with many leading figures on the far right for years, beginning in the mid-1970s with Pennsylvania Ku Klux

Klan leader Roy Frankhouser and Robert Miles, then a KKK leader and now leader of the neo-Nazi Aryan Nations. It was also during this period that LaRouche developed a close working relationship with one of the most influential thinkers on the far right, Willis Carto.

Carto is the founder of a variety of far-right enterprises including the *Spotlight,* a weekly tabloid newspaper with a paid circulation of over 115,000; Liberty Lobby, a Washington, D.C.–based organization; the Noontide Press, a publishing house; the Institute for Historical Review, a pseudo-scholarly organization devoted to proving that the Holocaust never occurred; and, as of 1984, a political party, the new Populist Party.

At the center of all of Carto's enterprises grows a virulent anti-Semitism, a passion which blossomed in the already-right-wing Carto after he met neo-Nazi author Francis Parker Yockey in 1960. Yockey had written a book, *Imperium,* which contained many of the themes that would be close to the hearts of far-right activists over the next three decades—in particular, the themes of the supposed cultural and biological superiority of the Aryan people and the need to rid America of the "Jewish threat." Carto published Yockey's book—which had gone out of print—through his Noontide Press and wrote a new introduction filled with anti-Semitic and racist invectives. Carto wrote, for example, that "Negro equality . . . is easier to believe in if there are no Negroes around to destroy the concept."

Like David Duke and other members of the far right seeking to gain access to the American mainstream, Carto understands it is necessary to speak in code to avoid shocking the general public. Carto claims he is an anti-Zionist, not an anti-Semite. However, a quick look at his book *Profiles in Populism* (a collection of *Spotlight* articles) proves that distinction meaningless.

In the book's glossary, Carto defines Zionism thus:

A secular conspiratorial scheme overtly aimed at ingathering Jews of the world to Israel but in reality a world political engine of massive power which effectively controls all aspects of Western political, intellectual, religious and cultural life. Zionism overlaps substantially into both capitalism and communism. Without Zionist support, neither capitalism nor communism could survive. Zionism is strongly antagonistic to all nationalism except Jewish nationalism.[17]

Sprinkled throughout the book are references to the natural superiority of the white race, dire warnings on the consequences of racial "blood mingling," denunciations of "the terminal insanity of contemporary American democracy," and even a quick sideswipe at feminism.

Like LaRouche, Carto has links with David Duke. Overlooked in the controversy arising out of the former Klansman's successful run, as a Republican, for a seat in the Louisiana legislature in early 1989, was the fact that just three months earlier Duke had received 44,000 votes as the presidential candidate of the Populist Party. It is not surprising that this information was rarely mentioned in media accounts of Duke's victory; few reporters had even heard of the party formed by Willis Carto in 1984 to serve as a political umbrella for "anti-Semitic, white supremacist forces . . . looking for a foothold in the political mainstream for a broad political agenda to turn the United States into a 'White Christian Republic.'"[18]

Since its founding, the Populist Party has drawn membership and fielded candidates from every major far-right group in the country, including the KKK, the Posse Comitatus, the American Nazi Party, the Farmer's Liberation Army, the National States Rights Party, the Christian-Patriots Defense League and the Aryan Nations. The Populist Party, which has chapters in 49 states, has also provided a base from which the Christian Identity movement can become involved in political organizing, and some leading Identity figures are also members of the Populist Party inner circle. The party's first chairman, Robert Weems, a former Mississippi Klan leader, is an Identity minister. And Colonel Jack Mohr, who has been described as "the traveling salesman of the Identity movement," is a member of the Populist's Speakers Bureau.[19]

Assuming, as did LaRouche, that desperate Midwestern farmers were ripe for the far right's message of fighting back against the "forces out there," the Populist Party targeted farmers early, and has enjoyed some of its most successful organizing among this group. Carto's *Spotlight* is read throughout the Midwest, and Populist literature calling for a federal policy to help "family farmers but not agri-business corporations or absentee owners" is handed out at farm auctions and farm protest rallies. One indication of the Populist Party's success organizing in the Heartland is the number of Midwesterners who subscribed to the now-defunct tabloid the *American POPULIST*. Eight out of the twelve states with the most subscribers are found in the Midwest.[20]

While the Populist Party passes itself off as a conservative party of "plaid shirts and polyester suits," the brown shirts and swastikas do show through from time to time. The party's agricultural policy attacks the "criminal Federal Reserve conspiracy" and "big international bankers"—favored far-right code words meaning Jews. The title of the Populist agricultural policy is "A Final Solution to the Problems of Agriculture," echoing Adolf Hitler's euphemistic term for the Nazi's systematic destruction of European Jewry.

Duke's Louisiana victory was a seminal victory for the Populist Party (and for the far-right movement in general) despite the fact that he ran as a Republican. As the party newsletter, *Populist Observer,* explained, "Party leaders . . . would naturally have preferred to see Duke run as a Populist instead of as a Populist-oriented Republican, but his close association . . . is bound to help spur party growth."[21]

In fact, as the Center for Democratic Renewal astutely pointed out in 1987—almost a year before Duke's victory—the Populists had adopted a "tri-partisan" election strategy, running as Populists where they could, and as Republicans or Democrats where necessary. Recalling the successful 1986 LaRouchian foray into the Illinois Democratic primaries, the CDR asked rhetorically in its *Monitor,* "Will David Duke take up where Lyndon LaRouche left off?"[22] The answer, quite obvious in the wake of Duke's victory, is yes. In one year Duke ran for office as a member of each of the three parties: as a Democrat in the presidential primaries, a Populist in the general election, and a Republican in Louisiana.

The "tri-partisan" path blazed by Lyndon LaRouche and paved by David Duke is sure to see increased traffic in the future. For all its irrational theories about eugenics, the fate of the lost tribes, and the cabal of international bankers, the far right is one of the most pragmatic political formations in the United States today when it comes to mass organizing. And so far that pragmatism has paid off. "In the last several years," says the CDR's Zeskind, "far-right groups have managed to establish a beachhead from which to spread their message of hate."

The success that the far right has had in organizing rural Americans recently is due to several factors. Just like Mussolini's Fascists in Italy in the 1920s and the National Socialists of the 1930s in

Germany, these latter-day totalitarians have taken advantage of the social and economic turmoil of their time and place. As rural communities started to collapse in the early 1980s, neither the Democrats nor the Republicans were responsive to the social and economic plight of these Americans.

But the far right was. Members of the Posse Comitatus and LaRouche representatives could be seen at farm auctions comforting families. While a smiling Ronald Reagan was on TV telling them that it was morning in America, the far right was confirming what rural people knew to be true: that their livelihoods, their families, their communities—their very lives—were falling apart.

The far right went a step further: it provided a detailed analysis of *why* rural communities were becoming rural ghettos. In a sense, it was less important what theories the far right offered than that its people cared enough to include these marginalized Americans in a broader political framework. In the eyes of many, the fact that Lyndon LaRouche and David Duke spoke to the very real problems that were ignored or even denied by mainstream leaders made them the only legitimate game in town. The far right gave the victims of the American nightmare of downward mobility the one thing they desperately needed—hope that the American dream could once again work for them.

The new rural poor were ready to follow almost any leader who offered them that hope, for while the day-to-day struggle to survive while mired in poverty is an embittering experience for anyone, the pain endured by this class of Americans—the inhabitants of the new rural ghetto—is in one way unique. Charles Silberman has rightly pointed out that "American cultural goals transcend class lines, while the means of achieving them do not."[23] But many of the new rural poor had not only shared American cultural goals—they had achieved them for a time. They had been *in* the middle class, *of* the middle class. They had tasted the good life and then had fallen from it. It is hard enough to watch from a distance as others eat, but it is an even more embittering experience to watch in hunger as others dine at the table where you once sat. The result is resentment and rage.

The far right understands rural peoples' alienation and exploits it, transforming their bitter desperation into political action that suits the right's own broader agenda. So what that their tangle of pseudo-legal procedures and quasi-religious doctrines is half-baked? So what that those who follow their advice end up either off the land or in jail—or dead like Art Kirk? The far right at least offers the possibility of salvation, and to the forgotten farmers and small-town residents that is sometimes enough.

Author James Coates suggests that many Americans have been "softened up" by the rhetoric of new-right leaders—especially by fundamentalist ministers Jimmy Swaggart, Jerry Falwell, and Pat Robertson—and made more susceptible to the similar but far more hateful message of the far right. Jerry Falwell's assessment that Jews are "spiritually blind and desperately in need of their Messiah and Savior," as well as the preacher's words to another audience, "A few of you here today don't like the Jew. And I know why. He can make more money accidentally than you can on purpose," are embryonic forms of the rabid anti-Semitism of the fanatic fringe.[24]

And clearly, Pat Robertson has added weight to the Identity call for a "Christian Republic" by saying, "The minute you turn [the Constitution] into the hands of non-Christian people and atheist people they can use it to destroy the very foundation of our society. And that's what's been happening."[25]

But CDR's Leonard Zeskind argues that Americans don't need a Pat Robertson or Jerry Falwell to soften them up with "soft-core" anti-Semitism. "There's already 1,000 years of Christian history on that," he explains. The connection between the far right and the new right isn't causal, rather they both arise from a preexisting and extensive body of anti-Jewish thought. As one theologian has pointed out, "It [is] clear that anti-Jewish ideology is much more deeply rooted in Christian preaching and even in some parts of the New Testament than had once been thought."[26]

We like to think of the Midwest as being above such crude passions as anti-Semitism and racism. Probably no region in the country has so benefited—and so suffered—under the process of mythicizing as has the fabled Heartland, where, as Garrison Keillor says, "every woman is strong, every man is good-looking, and every child is above average." We may laugh at that wholesome characterization, but we trust in its essential validity. Families in New York and Boston advertise in Midwestern papers for nannies for their children. A sperm bank on the east coast pays top dollar for Midwestern sperm because of the assumption that Heartland sperm is sure to be untainted by that "big-city" disease AIDS and, implicitly, by other kinds of corruption.

The Iowa-produced TV documentary "Harvesting Fear" reinforced this morally squeaky-clean stereotype when it portrayed farmers as the innocent victims of far-right (and out-of-state) leaders. "The very soul of the Heartland may be at stake," the report concluded, as if far-right groups and their ideology of hate are the spiritual equivalent of AIDS—an alien virus attacking an otherwise healthy Midwestern body politic.

"I know some minorities won't agree with me," a professor of political science at a small liberal arts college in Iowa recently told a reporter, "but on the whole, Iowans are more tolerant and more open than other parts of the country."[27]

When the state of Iowa recently acquired the house that appeared in the background of Grant Wood's painting *American Gothic,* one official proclaimed, "I think the *American Gothic* house is an image of Jeffersonian America. It says, 'This is the Midwest—where simple folk live and values are real.' "

But the Heartland has a dirty little secret. Beneath the warm smiles and bland platitudes about "simple folk" and "real values" lies the same racial and ethnic intolerance that blights American society elsewhere—and perhaps to an even greater extent. The far right didn't create bigotry in the Midwest; it didn't need to. It merely had to tap into the existing undercurrent of prejudice once this had been inflamed by widespread economic failure and social discontent.

"We set the Midwest apart as this little Camelot, but it's not," said a farm activist in Missouri. "We have a real problem with these things here, but we just don't want to believe it's true." The activist then asked that his name not appear with his words. "I don't want the Klan burning crosses on my lawn," he explained. He called back a half hour later. "You know when I said I didn't want you to use my name?" he asked. "I just wanted to make sure you didn't think I was joking. It's no joke."

The subtext of bigotry is so pervasive that small-town residents often don't even recognize its existence. A reporter in a small Midwestern city recalls being surprised at the level of racism he encountered in his hometown after a ten-year absence.

"I was in the service in the South," he says, "and saw a lot of racism there. But it wasn't until I came back here that I realized . . . whoa, rednecks don't just live in the South. People out here wouldn't know what you're talking about if you used the word 'Black' or 'African-American.' They still say 'nigger.' "

The reporter also asked that his name not be used.

The best recent assessment of the extent of anti-Semitism in the Midwest is a 1986 poll conducted by Louis Harris and Associates commissioned by the Anti-Defamation League.[28] In the poll of residents of Iowa and Nebraska, 27% of those questioned agreed with the statement "Farmers have always been exploited by international Jewish bankers, who are behind those who overcharge them for farm equipment or jack up the interest on their loans."

Also, 27% agreed that "Jews feel superior to other groups," and the same number said that Jews have too much power in this country.

The Harris poll found other distressing results. A 45% plurality of respondents over the age of 65 agreed with the statement that Jewish bankers were behind farm problems. And when presented with the statement that "Jews should stop complaining about what happened to them in Nazi Germany," only 48% of those polled rejected the statement outright (42% agreed).

In Nebraska, a 48% plurality agreed with the proposition that "when it comes between choosing between people and money, Jews will choose money." In Iowa, 39% of those polled agreed with the stereotype, while 44% disagreed with it. A large majority of people in both states over the age of 65, some 70%, accepted the statement.

The results of the Harris poll seem to indicate that ethnic intolerance is more widespread in the farm belt than elsewhere. In a 1985 national Harris poll, the same statement about Jews' choosing money over people was rejected by a margin of 63 to 30%.[29] But the CDR's Zeskind warns against reading too much into the results of the two surveys. What the Midwestern Harris poll really measured, he says, is the success of the far right's rural campaign over the last decade.

"You've got to remember that by 1986 these groups had been very active for seven years," he points out. "And until 1985 they had been pretty much unchallenged. I'm not sure that the level of racism is any higher in a small Midwestern town than in Yonkers, New York. Racism is a part of the overall American fabric."

While it is hard to quibble with that last assertion, one cannot explain away or minimize native Midwestern strains of intolerance. A high level of bigotry is neither alien nor new to the rural scene. Opposing the supposed foreign adulteration of "true American stock" by intermarriage was one of the central causes of the Country Life movement in the early days of this century.

Henry C. Wallace summed up this nativistic thinking succinctly—
if crudely—when he termed Midwestern farms "the breeding-
grounds of the Nation."

> The farms of the United States produce every year from a
> third to half a million more children than is necessary to
> maintain the farm population. These extra hundred thou-
> sands are sent to the cities every year, and the people left be-
> hind on the farm must proceed to their double task of feed-
> ing the cities both food and new blood. . . . The future of the
> cities of the United States seems eventually to lie in the qual-
> ity of the blood sent them from the farms. The native-born
> in the cities . . . are being replaced by the children of foreign-
> born parents and the children of the farms. Our greatest
> wealth is in the children of the next generation. The blood
> and education with which they are equipped determines in
> the long run whether our civilization is going up or down.[30]

Even earlier, during the waning days of the last century, the
original rural-based Populist movement contained a contradictory
mix of prodemocratic and antiblack sentiments.[31]

In fact, the attitude toward minority-group members has
changed very little in small town America since 1925 when sociologists Robert and Helen Lynd observed in their classic study of
Muncie, Indiana, "Jewish merchants mingle freely with other
business men in the smaller civic clubs, but there are no Jews in
Rotary; Jews are accepted socially with just enough qualifications
to make them aware that they do not entirely 'belong.' "[32]

This notion—still widespread in rural America—that Jews,
African-Americans, and other minority-group members "do not
entirely belong" is, in large part, responsible for rural people's
easy acceptance of the far right's agenda of hate.

That the far right hasn't made even greater inroads in rural
areas is thanks to the unflagging efforts of grass-roots organizations such as the Iowa-based Prairiefire and the Center for Democratic Renewal. These groups have been on the front lines in the
difficult and often dangerous battle against the racist and anti-
democratic ideas of the rural far-right movement—monitoring
hate-group activity, educating farmers about the totalitarian realities behind the far-right panaceas, and, perhaps most important,
providing an alternative political-economic analysis of rural

America's problems along with realistic policy suggestions for improving rural life.

But the efforts of these small and ill-funded groups go only so far. They are unable to fully counter the effects of the more numerous hate groups, and they are certainly in no position to bring prosperity to rural America. All activists have managed to do over the last few years is to keep the far-right movement in check. As the rural economy continues its slide, the beachhead established by the far right during the farm crisis of the 1980s will, in all likelihood, continue to grow in the 1990s. If genuine alternatives are not provided, a significant number of rural ghetto residents—bitter, desperate, and increasingly cut off from the nation's cities—are sure to seek their salvation in the politics of hate.

7

——

The Second Wave

*[Dying communities] constitute the ideological Achilles'
heel of laissez-faire capitalism.*

HARLAND PADFIELD
in *The Dying Community*

Columbus Junction, Iowa

When news leaked out that the old Rath packinghouse on the
outskirts of town was to be reopened, most residents of the
small river community felt like getting down on their knees and
shouting praise to the Almighty. Some, in fact, did just that.
Many believed that a bona fide miracle, a heavenly intercession
every bit as glorious as the burning bush—and far more practical—had occurred in their humble town.

After all, the 1,400 townspeople had been devastated when
the meatpacking plant closed in 1983, throwing almost 600
breadwinners out of work. Like any of a dozen other small
towns huddled in the southeast corner of Iowa, Columbus
Junction found itself tumbling into a sinkhole none of its residents had even suspected was there. But unlike those other
towns, Columbus Junction had now been singled out for salvation. Why? wondered residents. Better to not dwell on that.
Why had the Rath plant closed in the first place? Why did Job
have it so hard? No, one small statement is worth a thousand
big questions. Better just to endure the hard times and give
thanks for good ones.

And this certainly was the best of times. The new owner of
the plant, IBP (formerly called Iowa Beef Processors), was the

largest meatpacker in an industry of giants. It promised to employ 1,200 workers—double the number at Rath when it closed down. A hard-hit community in a hard-hit state was suddenly on the rebound.

"When people heard IBP was coming to town, they saw dollar signs everywhere," says Connie Lewis, who, along with her husband, owns both the local funeral home and the ambulance service.

It is not hard to understand that hallucination. The plant meant jobs and jobs meant paychecks—steady money for a change, money that would circulate up and down Main Street like an ocean wave washing into an estuary. From the workers to the grocery stores, and from there to the barber shops, cafes, pharmacy, appliance and clothing stores. Taxes paid by the new homeowners would improve city services, which had dwindled during hard times. The local school would benefit from the increased revenue, too. There were new textbooks to buy, sports programs to reinstate, computers to add. The children of Columbus Junction would have a shot at a brighter future.

On the day that IBP representatives and town officials formally announced the reopening of the plant, Robert Peterson, the chairman of the meatpacking company, separated a local banker from the festivities and quietly made him a promise: "Your town will never be the same."[1]

Two years later it's evident to even the casual observer that Peterson wasn't just boasting. Everyone agrees that Columbus Junction is today a very different place. From the traffic jams that clog Main Street on Friday nights to the new faces at the Hy-Vee supermarket, today's Columbus Junction is a different town. But the question that is debated down at the He Ain't Here Bar, or across the street at the barber shop, or at church, or anywhere two or more longtime residents happen to cross paths, the question they chew over and that often divides them is, Is Columbus Junction a better town?

"There are very few people in town who don't have their minds made up about IBP and what it's done to the town," says Keith Isley, the pleasant young news editor of the local weekly newspaper, the *Columbus Gazette*. Taking shelter in the newspaper's basement office from an oven-like August sun, Isley is himself extremely cautious about saying anything that can be taken as either an endorsement or a condemnation of IBP. Asked specifically whether he thinks IBP's presence is basically good or bad for the community, Isley is obviously put off by the question. Like most journalists, he would rather ask hard questions than have to answer them. He smiles nervously, stroking his neatly trimmed goatee as he frames an answer. "That's a difficult question," he says. "It's like a diamond that changes facets as you turn it."

The most positive facet is the undeniable fact that IBP has brought jobs into a community that desperately needed them. And yet, little else has worked out as planned. The wave of money that was supposed to flow around town as a result of those jobs apparently stalled far out at sea. In fact, retail sales in town actually declined in the year following IBP's arrival. Many blame IBP's policy of paying its workers a starting wage of $6 an hour—just half of what Rath paid in 1983, not adjusting for inflation—for aborting the expected wave of prosperity.

The wave that *has* hit town was completely unexpected. It is a steady stream of people—poor, unemployed people, sometimes couples, but more often single men 18 or 19 years old drawn to Columbus Junction from Tucson, Seattle, Oklahoma City, San Diego, and a hundred other cities by the promise of a job. This influx of poor transients has created a housing crunch in Columbus Junction. Those who can afford the cost live in motels, often doubling and tripling up to pay for a small room. Those without money sleep under bridges while waiting for a position at the packinghouse to open up. Many IBP employees continue to live under open skies or in their cars even after beginning work. At $6 an hour it takes some time to scrape together a month's rent and damage deposit.

In the summer, whole families live in their rusty cars parked at a nearby river boat landing, existing on canned goods and on fish caught with poles cut from willow branches. At night, the roadside ditches in the countryside surrounding Columbus Junction are dotted by the orange glow of cooking fires where small knots of men gather to prepare their evening meal.

"When we pass an abandoned out-of-county car along the road," says resident Connie Lewis, "my husband and I say, 'There's another IBP worker.' "

In response to the housing crisis created by IBP's arrival, the company bought ten mobile homes in the nearby town of Conesville, and rents them out, unfurnished, to employees. The rent—which runs between $175 and $275 a month, not including utilities—is deducted from the employee's paycheck. Butch Bennett, an ex-Rath worker and now owner of the Sportsman's Inn bar in Conesville, is disgusted by the practice.

"They just brought in a group from Las Cruces, New Mexico," says Bennett. "Most of them can't speak English. IBP

promises them these great jobs. They deduct the travel
expenses from their paycheck. Then they deduct rent from
their paycheck. And then these people are left with nothing.
You know that old song: 'Fifteen tons and what do you get?
Another day older and deeper in debt.' Well, that's what's happening to these people. They owe their souls to IBP."

On the far north edge of Columbus Junction, just at the
point where Main Street dissolves into a dusty gravel road,
stands a weather-beaten two-story building that until recently
housed the offices of the town chiropractor. Despite its run-
down appearance, some in town believe this building symbolizes
the kind of creative spirit that will keep towns like Columbus
Junction strong, no matter what changes are in store. The building has been converted by a local businessman into temporary
"sleeping rooms"—8 by 10-foot cubicles used by IBP workers as
little more than a spot to lie down in after work. For that privilege the workers pay $60 a week plus utilities. The accommoda-
tions are offered on a temporary basis only; the workers must
make other plans after one month. They drift in and out of the
dingy rooms, one replacing the other with the predictability of
an assembly line, and they leave behind no record of their passing aside from the odd bits of graffiti carved into the walls.

Mark, a 30-year-old IBP employee who has been living in one
of the sleeping rooms for two weeks, isn't sure where he'll go
when his time here runs out. "What I'd really like," he confides,
"is to find a nice little apartment close by that we could move
into. But how likely do you think that is?"

In addition to himself, "we" refers to his wife and two small
children, who now live in Waterloo, 200 miles northwest of Columbus Junction. Mark usually goes home on weekends, but he
recently hurt his back and now can't drive that far. He continues to work, however, despite the pain, saying that his family
counts on his paycheck. Mark works on the cutting-room floor,
hooking hog stomachs onto slots at the rate of 1,020 stomachs
an hour. He earns $6.10 an hour.

Mark sits on a brown couch that reeks of cigarette smoke in
what was once the chiropractor's waiting room. It is a long narrow room containing, in addition to the couch, a broken reclin-
ing chair, a pay phone, and a small television set bolted to the
wall.

Down a narrow corridor leading from this lounge are the
sleeping rooms. We hear the muffled sound of a door opening
and then closing again somewhere along the corridor, and a
few seconds later an obese young man wearing black gym

shorts and a green T-shirt walks into the room. He has the gait of an old man: slow and careful. He sits down on the couch next to Mark and begins eating his breakfast—three packages of Reese's Peanut Butter Cups and a can of Pepsi. The whites of his eyes, nearly hidden behind puffy lids, are shot through with bright red lines.

"You seen my hat?" the new arrival asks Mark between bites of candy.

"What's it look like, John?"

The younger man shuts his eyes for a moment while he thinks of his missing hat and then announces, "I think it's blue and it says 'Same shit, Different day.' "

"No, I haven't seen it."

"Oh. I was pretty fucked up last night," John says in a monotone, and then gives up on conversation, turning his attention to a TV game show while polishing off breakfast.

Mark returns to talking about IBP. He doesn't expect to work there much longer, he says. He hopes to get hired back at John Deere, the farm implement manufacturer, in Waterloo in just a few months.

"I made good money there," he says, a large smile spreading across his face at the memory. "Fifteen dollars an hour back in 1980. Of course, I was union then. I'm for unions. Here, forget about it. My supervisor called me nigger and I complained." He shrugs. "Nobody did anything."

"I don't like unions," interjects John in the same flat voice he used before. "The government already takes enough money out of your paycheck. I don't see what no union would do."

"They'd make them pay us more."

John shrugs. He says that he likes his job "robbing" neck bones—cutting the meat off of them with an electric Wizard knife. He is a farm kid from up north. The light-brown fuzz just appearing on his cheeks seems at odds with the vast fatigue and utter acceptance of his lot that hangs over him. Asked if he likes the townspeople, he says, "I don't really see too many of them. I party with the people at work. I like the bars, though. The people are nice to you."

Mark shakes his head and laughs. "They should be, man. They're making a killing off of us—just like IBP," he says bitterly. "They're all taking advantage of us here."

John drains the last of his Pepsi and drops the empty can to the floor, where it falls on its side and clatters out of sight beneath the couch.

"I don't care," he says to no one in particular, his eyes fixed on the TV screen, his jaw slowly working on another piece of candy.

For their part, local residents are having a hard time getting used to the seemingly endless parade of Marks and Johns passing through town. According to *Gazette* editor Isley, it is common for 50 to 60 IBP employees (at the 1,200-worker plant) to be replaced each week—giving the packinghouse an annual turnover rate of 240%. Change on that scale has had a profound effect on a small town where many adults still live in the houses they were born in.

"Iowa has always had a unique sense of stability," explains Isley. "That's threatened when you have people moving into town for a couple of weeks and then moving out. You lose a sense of community."

The constant turnover has also put a strain on community resources already stretched thin by the failing farm economy. Even the local police department is feeling the squeeze. The number of criminal cases in Columbus Junction jumped from 167 in 1986 to 290 the following year.

"They [IBP workers] get a job for a couple, three weeks, write a few hundred-dollar checks, then they leave," explains Police Chief Ernest Whiting.[2]

But it isn't just an increase in the number of rubber checks bouncing around town that concerns residents. "Big-city" violence has come to Columbus Junction. Fights regularly break out in formerly quiet bars and spill out onto the streets. The level of violence at the IBP trailer court in Conesville has become so great that local ambulance drivers now refuse to respond to calls there without police backup.

"We're just seeing so many problems we haven't seen before," says Connie Lewis, owner of the ambulance company. "It's pretty scary when you get a call to help someone who was slashed in the back with a switchblade."

But the incidents of violence in town are only hints of the dangers found inside the plant itself. Workers at IBP packinghouses around the state—there are four in Iowa—tell of accidents that scar, maim, and cripple with alarming frequency—grisly stories that could have come from Upton Sinclair's 1906 exposé of the meatpacking industry, *The Jungle*.

Take, for example, what happened in November of 1983 to Janet Henrichs, an employee at IBP's Storm Lake, Iowa, plant. Henrichs, an 18-year-old high school dropout, was working at a machine which stripped the skin off pork shoulders, when the glove on her right hand became entangled in the machine's "feed" mechanism. In seconds the device had completely skinned her hand, in the process mangling tendons, ligaments, and muscles all the way up to her wrist. Henrichs was left permanently disabled.[3]

Although packinghouses are by their nature dangerous places, a court later ruled that in Henrichs's case negligence on the part of IBP was to blame for the accident—not problems inherent in the industry. It was found, for example, that although the skinning machine came equipped with an emergency foot-operated "kill switch," that feature had been removed. There were other irregularities: At the time of the accident, the equipment had been set to run three times faster than the manufacturer's recommended speed. And a sign specifically warning operators not to wear gloves while operating the machine had been removed. The missing sign had also stressed the importance for anyone operating the machine of first reading the owner's manual. Henrichs had never been shown the manual.

Former IBP employees and union officials charge that what happened to Henrichs is just the tip of the iceberg, an accusation that was given credence in August 1987, when the United States government proposed a record $2.6 million fine against IBP for allegedly trying to cover up 1,038 job-related injuries and illnesses at its Dakota City, Nebraska, plant.[4] According to the government, the company had kept two sets of accident reports: one for itself and one—with far fewer injuries listed—to show to government safety inspectors. A year later the government proposed an even larger fine against IBP, $3.1 million, this time for willfully ignoring a hazard that was crippling hundreds of workers. IBP employees were developing a painful and disabling wrist condition called carpal tunnel syndrome, a malady caused by the repetitive motions made in trimming animal carcasses. The government alleged that although IBP knew that hundreds of its employees were developing the condition, the company took no action. A government investigation determined that more than 600 workers at just one IBP plant showed symptoms of the disorder. When one worker complained to his supervisor that he was developing knots in his wrist and that his hands were growing numb (both symptoms of carpal tunnel syndrome), he claims he was told, "Stand up and be a man, quit being a puss. Don't you want to work here anymore?"[5]

Such revelations moved Democratic presidential candidate Bruce Babbit to dub IBP a "corporate outlaw" during the 1988 presidential primaries. The company, Babbit said, was "a monu-

ment to everything shabby and backwards and wrong in the
American economy."[6]

Many in Columbus Junction today agree with Babbit's characterization, while others maintain the hope that despite all the
initial problems, IBP will turn out to be an asset to the community

It's hard to believe today, given the climate of contentious-
ness that now permeates the small town, but back in 1986 there
was little such debate. The company was simply a lifeline tossed
to a community falling into an abyss. Sure, they will tell you, a
few townspeople may have heard some unsettling reports about
IBP, but the community, already hurtling into darkness, was
not about to ask too many hard questions of the company offering the only lifeline within reach. After all, the locals will tell
you in a phrase that is fast becoming the state's unofficial motto,
beggars can't be choosers.

"It was a simple question then," says Connie Lewis. "Would
we rather have a company with some problems or an empty
building? That was our choice. I can't say whether it was a good
one or a bad one, but back then it was our only one."

Sometime in the mid-1980s, when it became obvious to even the
staunchest supporters of President Reagan that the farm crisis
was not merely a red herring cooked up by frustrated Democrats
and that the rural communities supported by farmers were indeed in desperate straits, an old term was given new life: "rural
economic development."

Today, there are frequent conferences on the subject, which is
also referred to simply as rural development. Countless newspaper and magazine pieces as well as dozens of scholarly articles
have been written on it. Everyone from the governor of Iowa to
the waitress at the corner cafe in Columbus Junction has something to say about rural development. And broadly speaking, the
consensus is, We need it, and as much of it as we can get.

Rural development is seen as a panacea for what ails our small
towns. But while almost everyone sings the praises of rural development, there are few capable of defining exactly what it is they
mean when they use the term. Pressed on the point, many a state
legislator may say rural development means jobs. But development is more than jobs, as the residents of Columbus Junction
discovered. An ironic result of this confusion is that the very
factor on which the inhabitants of rural ghettos have pinned their

hopes—a new industry moving into their area—often ends up entrenching and institutionalizing poverty and its attendant ills. Considering all the damage done in its name, rural development, as it is commonly practiced, might well be considered the second wave of the farm crisis.

Given the outmigration of much of the middle class from rural ghettos, it is not surprising that the companies moving into these areas are labor-intensive industries which pay low wages. Indeed, these companies *look for* just such communities, where land values are depressed and cheap labor is available. After all, what other industries can make use of a largely uneducated, mostly unskilled work force? Once in place, these companies attract more uneducated, unskilled, and unemployed workers to the area, furthering the ghettoization process by increasing the ranks of the available labor pool and acting to keep wages depressed. The wave of poor immigrants also makes the area less attractive to the remaining middle-class residents and so increases their exodus, again adding to the downward spiral.

Guiding much of current rural development is the notion that state and federal aid should focus on helping only a few "growth centers," leaving the bulk of the Heartland's small towns to fend for themselves. According to this position, it is foolish to waste scarce resources on the residents of dying towns; aid should be funneled only to those communities that already show promise. Advocates of this idea attempt to make the notion more palatable by implying that the inhabitants of rural ghettos have only themselves to blame for their continuing troubles. A University of Nebraska at Omaha report on rural development took this tack when it stated that "those [towns] that are likely to succeed will be willing to commit the extra effort to secure the resources necessary to reach a minimum level of capacity." Left unstated was the implication that "unsuccessful" rural communities just didn't try hard enough. The report reinforced this perspective when it concluded, "The old adage, 'There's no point in helping those who can't, or won't, help themselves,' is clearly true for local governments."[7]

While a recent USDA report on rural development forsakes the "blaming-the-victim" tone evident in the Nebraska study, the federal agency takes a position even more damaging to troubled rural communities when it asserts that a primary purpose of rural development should be to facilitate "a smooth and rapid movement of capital and labor from ... less to more competitive locations."[8] Some rural development experts have derided the policy advocated by the USDA as being "just short of triage."[9]

In addition to a bias against helping the neediest towns, current rural development policy has several other flaws. In a desperate attempt to bring jobs to their devastated regions—and then to keep them there—state and local governments attempt to outdo each other in making concessions to manufacturers, offering businesses a variety of inducements such as tax breaks, job training programs, and development bond financing. In a skirmish typical of this "new war between the states," South Dakota in 1989 offered a low-interest $700,000 loan to a growing computer company located ten miles over the state line in Iowa. Iowa's governor intervened, offering the company a forgivable loan of $300,000 if it would stay in the Hawkeye State. But the fact that South Dakota had lower worker's compensation and unemployment costs, as well as lower taxes, proved too sweet a deal to pass up; the firm moved.[10] A study of the rural development policies of six Midwestern states concluded that "while the quest for competitiveness may be global, the states . . . find themselves primarily in competition with each other."[11]

This cutthroat competition for jobs has led to a situation in which those conditions that are the very worst for their citizens—low wages, antilabor laws, inadequate environmental protections—are exactly what states promote. An economic development plan crafted for one section of Iowa contains a prime example of this attitude. In a section listing the advantages of locating in Iowa, the plan boasts, "Iowa's annual average pay of $15,540 was the 10th lowest in the nation." On the same page, the authors of this plan observe, "Strikers are not eligible for benefits [food stamps and other social service aid], which is a very positive feature from the perspective of business."[12]

There is nothing original about this mentality. This "new war between the states" actually began almost immediately upon the conclusion of the Civil War. In an attempt to make the postbellum South more appealing to Northern manufacturers, for example, the Alabama legislature repealed the state's ban on child labor. Other Southern states, not wanting to be left behind, followed suit. The lure of cheap labor—including child labor—attracted much of the North's textile industry.

The battle in the South took a decisive turn in 1936—a change that remains the leitmotif of rural development in all parts of the country to this day. In that year Mississippi became the first state to begin a regular program of granting subsidies to industries promising to locate inside their borders. The program, called Balance Agriculture with Industry (BAWI), heralded the modern era of direct state-run competition for industrial development.

Previously, the role of offering incentives to industry had, by and large, been left to individual communities. Towns commonly floated bonds to finance the construction of new factories and warehouses, and handed the buildings over to the manufacturers or rented them out for a few dollars a year. Abuse of such programs was widespread. Desperate communities often made rash promises to manufacturers, agreeing to deals that compromised—and sometimes negated—any benefits that the new jobs would have provided. One such questionable practice was the institution of salary deduction plans. Under these plans, workers signed contracts allowing their employers to deduct from 5 to 7% of their (already meager) paychecks each week. The money was used to pay for building or renovating factories.

An even more glaring example of worker exploitation was the diversion of WPA money to build a "vocational school" in Ellisville, Mississippi, in 1935. The "school" was actually a hosiery factory run by a Pennsylvania firm, the Vertex Hosiery Company. Vertex operated the plant rent-free, and since the plant was technically a school, it paid no taxes. "Students" at the Vertex "school" worked 40-hour weeks and were paid wages half the standard rate. At the end of their training period, these exploited workers were supposed to be promoted and paid the normal factory wage, but this rarely happened. More often, trained workers were let go and replaced by more "students."[13]

The BAWI program standardized the process of state industrial recruitment and eliminated the worst abuses of the system. Still, there were several drawbacks to the program. BAWI institutionalized the already-antiunion sentiments of the South. States and communities entered into a Faustian bargain with manufacturers, and said in effect, You give us jobs, and we'll put all the powers of government behind your efforts to prevent unionization.

A sample BAWI contract read:

The Second Party (the company) pledges itself to be fair in all of their dealings with employees and to pay fair and reasonable wages, and the First Party (the city) agrees that it

will so far as possible prevent any interference from outside
sources which may cause or result in labor disputes or trouble
and the payroll guarantee hereunder by the Second Party
shall be cancelled during the period of any labor disturbances
caused by outside interference.[14]

As a result, BAWI not only attracted manufacturers who paid
low wages but it undercut the possibility of workers organizing to
bargain collectively, the only action capable of increasing workers'
leverage with their employers in order to improve working conditions.

It is uncertain how many of the companies moving into Mississippi were attracted by BAWI incentives and how many would
have moved south even without such a program. Although backers of the program pointed to an impressive array of statistics
showing a rise in manufacturing jobs under BAWI, other Southern states which didn't have a similar program enjoyed even
greater growth rates during the same period, raising doubts about
the effectiveness of BAWI in industrial recruitment.[15]

It is questionable whether today's attempts at "smokestack chasing" are any more successful than were those made by Mississippi
under BAWI. Regardless, industrial recruitment is not the answer
to rural America's problems. For one thing, there are simply not
enough manufacturing plants to go around. In 1986 there were
some 25,000 towns across the nation competing for approximately
500 major industrial relocations.[16] Neither is manufacturing expected to be a major source of new jobs in this country in the
coming era. But for all its drawbacks, industrial recruitment continues to be the primary mode of rural development in the country today, and states remain the basic agents of that policy. The
war between the states continues, with increasing sophistication—
and, often, with increasing social destruction. A state's "business
climate" is now determined by a wide array of concerns that include taxation (hopefully low or nil), union power (little or none),
environmental regulation (slight or unenforced), and—perhaps
most important—the level of the "social wage," that is, the degree
to which the state shelters its workers from insecurity through a
variety of social welfare programs including food stamps, AFDC,
unemployment insurance, and others. From this Alice-in-
Wonderland perspective, the less stable the community, the better
positioned it is for development.

The results of following such a path are predictable. While the South is consistently ranked as having the best business climate in the country, at the same time it is ranked, just as consistently, as having the worst record in support of education, wages, and environmental protection. In 1980, the South contained 66 of the country's 75 most industrialized counties and 61 of its 75 poorest ones.[17]

The cost of winning the battle for industrial recruitment is best seen in Tennessee's successful fight to attract what had been called the "plum of the decade," General Motors' Saturn plant. The auto maker said the sprawling complex, heralded as the largest one-time investment in U.S. history, would provide jobs for 6,000 workers—with perhaps another 15,000 jobs somewhere "down the road." The company announced it would pour $3.5 billion into the plant.[18]

Within days of the announcement, some 20 states were trying to entice General Motors into locating the new plant within their respective borders. It was an amazing spectacle: schoolchildren wrote plaintive letters to GM executives, governors came calling like eager suitors to woo the girl of their dreams with fresh-cut bouquets of tax incentives and whispered promises of weak unions.

In June of 1985, GM announced that it had selected the rural community of Spring Hill, Tennessee (population 1,275), as the site of the Saturn plant. According to Spring Hill Mayor George Jones—who heard the news on the same day the rest of the nation did—the announcement "was like a star falling out of the sky."

After obtaining a multi-million-dollar package of concessions from state and local governments in the form of tax deferments, reduced utilities charges, and highway construction projects, GM announced a change in the plan: there would only be 3,000 jobs—half the original prediction—and the $3.5 billion investment was to be scaled down accordingly to $1.75 billion.

Four months later, Tennesseans received another surprise: the company revealed that most of the jobs at the plant would not go to local residents. They had already been promised to the members of GM-UAW units across the country.

As in the case of Columbus Junction after IBP's arrival, the coming of Saturn to Spring Hill also attracted a steady stream of the unemployed, many of them destitute and all of them hoping to get work at the plant. The city, unable to care for this influx, passed the problem along by buying them one-way bus tickets to nearby Nashville.

Some residents benefited from skyrocketing land values that came in the wake of Saturn. "We didn't have any real estate offices in Spring Hill before Saturn hit," said a local bank manager. "The next day we had five. . . . Land speculators blew into town with money in hand, bought land, and flipped it."[19] Land that sold for $1,000 an acre in July was fetching from $5,000 to $10,000 an acre (and as high as $35,000 an acre) in September. Of course, homeowners could only gain from the situation if they were to sell their homes and leave the only community many of them had ever known.

And what was the effect of rapid economic growth (as measured in this case by a dramatic increase in property values) on the less fortunate residents of Spring Hill—on those who could not afford to own their own homes? The many families who rented houses and apartments weren't confronted by the hard decision facing landowners. Unable to afford the rapidly rising rents that resulted from escalating land values, they had no choice but to move on.

A 1988 study of the situation in Spring Hill concluded that while it was not yet possible to make a complete assessment of the effects of Saturn's coming on the community, "What is known is that the three reasons to do industrial recruitment—jobs for locals, a better tax base, and community improvement and an era of progress for those who reside there—have not materialized. Many long-term residents have little hope they ever will."[20]

Clearly, many of the benefits linked to industrial recruitment are exaggerated or simply nonexistent. When proponents of this form of rural development make their case, they often cite the multiplier effect. This principle states that for every manufacturing job attracted to a town, a certain number of nonmanufacturing jobs are also created in a spin-off effect. The U.S. Chamber of Commerce puts the multiplier effect at 3.0—meaning that three nonmanufacturing jobs will be created for every one new position in manufacturing. But the actual numbers are far less heartening. The average multiplier has been determined to be 0.3, which means that rural communities can expect that one new job will be created around town for every three new manufacturing positions attracted to their community.[21]

But worse than simply overestimating the benefits of industrialization are the negative impacts of this kind of economic development, impacts that are often masked by the way in which we measure economic change. Probably the most serious mistake made in assessing the benefits of rural development strategies is to use "economic growth" as a yardstick for measuring success. Few standards are more misleading. To determine whether a change is beneficial or harmful to a community, we must look at how that change affects each segment of the population individually. Generally, those who stand to benefit by the new business are those who are already well-off: bankers, real estate owners, and other holders of property and wealth. As a group, the poor benefit least. In fact, the poor, many of whom live on fixed incomes, are often hurt by such a change; for while the prices of goods around town rise as money comes into the community, the incomes of the poor, especially the incomes of the elderly living on Social Security, remain the same.

That boomtowns are a bust for many of their citizens is apparent from even a cursory look at the country's latest success story, the Sun Belt. Even before the region started breaking down into a few isolated "sunspots," the distribution of incomes within the area showed a growing inequality. The wealthiest 5% of the Sun Belt population had a larger share of income (16.4%) than in any other region in the country. And the bottom 20% had a smaller piece of the income pie (4.8%) than did a similar population anywhere else in America.[22]

If attracting new industry is held to be the crowning achievement of "rural development," then the jewel in the crown is the successful wooing of high-technology industry. High-tech represents progress, say its backers. The future, they say, *is* high-tech. It is profitable, environmentally clean, and wide open for development. The movers and the shakers in rural states look out over corn and bean fields and in their mind's eye they see the next Silicon Valley. They see themselves as modern-day Jasons, sowing tax abatements, antiunion laws, and no-interest start-up loans in lieu of dragon's teeth, and then standing back as shiny high-tech factories pop up from the furrows like the Colchian soldiers. But just like the magical soldiers in the Jason myth, high-tech industries have a habit of turning on their creators.

Part of the problem is with language. The term "high technology" is too fuzzy. Just like "rural development," "high-tech" has no precise meaning, a problem that was underscored in 1988 when Eastman Kodak announced it had chosen Cedar Rapids, Iowa, as the site for its $50 million biotechnology plant. State development officials were thrilled. One told a reporter that the Kodak plant was exactly the kind of high-tech industry they were hoping to attract. Iowa Governor Terry Branstad, speaking of the new plant at a luncheon sponsored by the Iowa Future Farmers of America, was just as enthusiastic.

"What happens in biotechnology in the next decade might well surpass what happened in micro-electronics in the past," he said.[23] In reporting the governor's talk, one newspaper boiled Branstad's message down to the essentials in its headline: "Biotech promises bright future."[24]

Everyone seemed agreed that the Kodak plant was just the first of many such high-tech projects that would save the state.

Or almost everyone.

"The Kodak plant is nothing *I* would want to plan *my* economic future around," says Mary Bruns, who was researching Kodak's location decision as a graduate student in urban and regional planning at the University of Iowa. In Bruns's report on local economic development policy, she concluded that there was far less to the Kodak plant than had met the eye of the development crowd. Not only was it unlikely that the plant would form the nucleus of a new biotech industry in the state, she wrote, but the majority of jobs at the Kodak plant itself would not live up to the "high-tech" reputation. The problem was with the word "biotechnology."

Biotechnology is simply the use of living organisms or their components to make or modify products, a technology that includes fermentation and that has been practiced for centuries. The "new" biotechnology—what most people think of when they hear the term—grew out of scientific discoveries of the 1950s involving the structure of DNA. Within the new biotechnology, Iowa development specialists failed to distinguish between "genetic engineering," which is research- and labor-intensive, and "bioprocessing," which is highly automated and more capital-intensive. The Kodak plant is a bioprocessing center, devoted to manufacturing, not research. While biotechnology industries engaged in genetic engineering are what Iowa (and most other states) hope to attract, those firms usually locate near urban centers on either coast, where there is a large supply of scientifically skilled workers, more venture capital, and an existing academic environment that facilitates scientific entrepreneurship. Given this background, rather than being the first of many research-oriented high-tech plants to grow in the fields of Iowa, the Kodak facility may, at best, be the harbinger of more bioprocessing manufacturing plants. The jobs offered by such plants may also be below the expectations of many for whom high-tech means high pay. On this point Bruns wrote:

> At first, it appeared that the plant would offer employment opportunities for professionals graduating from Iowa's universities. Such opportunities would certainly alleviate the exodus of Iowa's graduates from the state. However, Iowa's universities do not offer strong professional programs in the areas needed by Kodak, and the company will apparently recruit professionals on the national level. . . . The types of jobs that will be offered at the Kodak plant are more mid-level, technician-type positions.[25]

The high-tech field as a whole, not just bioprocessing, has an undeserved reputation as a provider of good jobs. "Despite the image of highly-skilled, highly-educated employees," wrote Marc Miller, senior editor of *Technology Review,*

> high technology actually promotes a two-tiered work force: the elite professionals and the production workers. Two out of three (65 percent) employees in high-tech firms work in low-paying production, clerical, and technical jobs. Instead of advancing up the income scale, the average semiconductor production worker receives an hourly wage that is only 57 percent of the rate paid his or her counterpart in the typical unionized steel or auto plant.[26]

Nor is the crème de la crème of the high-tech world—the research park—immune from the normal vicissitudes of the business world. This segment of the highly competitive industry is, in fact, particularly risky. Approximately half of all the research parks which opened to optimistic pronouncements in the last several years have already closed. Another 25% have been forced to incorporate manufacturing plants and office space as a means of surviving.[27]

On the opposite end of the spectrum from the shimmering—albeit mostly illusory—promise of high technology there is another trend afoot, the reemergence of a mode of production that was already widespread in the nineteenth century: industrial homework. An increasing number of rural women are finding that industrial homework—from embroidering chic jogging suits to assembling stuffed toys in their homes—is the only work around. Advocates of homework say that the growth of this industry is a boon for rural women. Homework, they maintain, allows for employee flexibility in balancing the demands of work and family, increases worker autonomy, and protects the right of Americans to choose where they work.

But in fact, industrial homework is becoming to today's rural ghettos what sweatshops were to the immigrant tenements in the first half of this century: vehicles for the exploitation of a mostly female work force, characterized by low wages, unsafe working conditions, child labor, and little or no government regulation.

Franklin Roosevelt's New Dealers believed that the low wages paid to homeworkers and the competition homework presented to factory employment hampered their efforts to revive the post-Depression economy. In 1942, the National Recovery Administration banned most forms of industrial homework. In 1981, under President Reagan, the Department of Labor (DOL) sought to drop all restrictions on homework, but lost a court challenge mounted by the International Ladies Garment Workers Union (ILGWU).

The union victory was short-lived. The Administration succeeded in lifting the ban on knitted outerwear in 1983 by agreeing to a registration process in which employees were required to keep records on wages paid to homeworkers. The move was a "demonstration project" of sorts. The government hoped the experiment would prove it was possible to allow the wide-scale re-institution of homework while preventing the abuses traditionally associated with the industry.

Although a DOL study found that 75% of the employees in the newly deregulated sector were guilty of wage-and-hour violations, the government continued to lift restrictions.[28] In January 1989, the DOL announced its intention to drop the last remaining homework ban—on women's apparel.

While debate over the legalization of homework has focused almost exclusively on its urban implications, rural women are just as vulnerable to exploitation as are immigrants in New York's Chinatown, an area notorious for homework abuses. The two groups—urban immigrants and rural women—have far more in common than at first meets the eye. Most importantly, they share lives circumscribed by poverty. The poverty rate in rural ghettos is now almost equal to that found in inner-city areas. And because rural families are deeply attached to their community and their land (which in many cases has been in the same family for several generations), moving to a better labor market is not an attractive option.

Homework has added appeal in rural America because women's roles there tend to be more narrowly defined and more rigidly enforced than in urban areas. The idea of a woman working for pay outside of the home, especially if she has children, still meets with great opposition in Heartland communities. Homework is seen as an acceptable way for these women to enter the paid labor market without violating social norms—for example, without asking men to share in the responsibility of child rearing. Industrial homework thus forces women to do two jobs, squeezing homework in between the traditional female rural chores of farm work, child rearing and housekeeping. Homework allows society to conveniently ignore two extremely important—and converging—issues of our day: the devaluing of labor (especially rural labor), which has forced more women to enter the job market to make the family's ends meet, and the assumption that child rearing is "women's work," that is, not worthy of financial compensation.

As Eileen Boris, assistant professor of history at Howard University, said in an interview, homework

> encourages the view that women are only secondary earners who need not have jobs that pay better, and who only want to work a few hours a day. Letting women take in homework in itself hardly solves the underlying problems of why women choose homework: women's nearly exclusive responsibility for care of dependents, a sex-segmented labor market where women's work earns about 60 percent of men's, and the undervaluing of women's labor in both family and market.
>
> Women's place within the family, a factor that encouraged homework in the origins of industrialization, persists as a reason for its continued attractiveness, but also reinforces the very sex-segmentation behind the system. . . .

Connie Jorgenson, a middle-aged resident of Clarinda, Iowa, a small town in the southwestern corner of the state, was a homeworker from 1981 until 1985, sewing appliqués on sweatsuits at piece-rate pay for an Iowa-based company called Bordeaux. Like other residents of small rural communities that were hit hard by the farm crisis, at first Jorgenson felt lucky to have found work that allowed her to earn some money and still remain at home to care for her two children.

Sitting at a kitchen table stacked high with sweatshirts that she now appliqués as a self-employed seamstress, Jorgenson says that it didn't take long for her to sour on her job with Bordeaux.

"At first you just think about the work you're doing," she says. "You don't think about things like how much you're making an hour. But when I sat down and totaled it up, I was shocked. The most I ever earned during my years sewing for them was $2.25 an hour. And in one year I averaged around $1.50 an hour. But what could I do? We needed the money—what little there was."

Jorgenson often had to recruit her family to help out, especially when deadlines were approaching. Her husband would some-times stay up all night cleaning finished sweatsuits while she sewed. The couple's 16-year-old son helped cut out designs and fold clothes. Even their eight-year-old daughter was put to work clipping threads.

Child labor has long been associated with industrial homework. In fact, when reformers campaigned against homework in the early 1900s, it was the widespread exploitation of children under the system that they decried most loudly, often comparing the hard lives of homeworking minors to those of the "workhouse" children in Charles Dickens's novels. The problem remains today. As Allen Meier, Iowa commissioner of labor, points out:

> Once you get a punch press or an industrial sewing machine in the house, how do you keep a 12-year-old away from it? Say he wants a bicycle. You let him work on the machine for maybe a couple of hours each weekend—to earn money for the bike. After a while, the kid is bound to think, "Hey, if I work every night, all night for a week, I'll get the bike right away." And unless a kid shows up in school with some fingers missing someday, no one is going to ever know that they're doing industrial homework.

In an investigation of homework in Rhode Island, the chief labor standards examiner for the state reported that some home-workers "keep children up until all hours of the night, doing very simple tasks with jewelry. In one home, we found a child in a high chair putting backs on earrings."[29]

The temptation to use children "just to help out" is overwhelming when a homeworker is struggling to complete her work on time. It is, after all, for the survival of the family that most women turn to homework in the first place. But where exactly does "helping Mom out" end and child exploitation begin? There are no clear dividing lines, and so in a very short time children become indispensable parts of the homework "team." What we rejected in our factories over a century ago, we are today willing to allow into our homes—or at least into the homes of the rural poor.[30]

There are also health hazards associated with homework. Connie Jorgenson developed a painful condition known as "sewer's neck" after sewing for several hours in a row. Others warn of the dangers posed by the unregulated exposure to hazardous chemicals in the form of glues and solvents.

For manufacturers, there are many benefits to using homeworkers. By replacing factory workers with homeworkers, employers are able to eliminate many of their overhead costs of work space, employee facilities, and benefits. Hiring homeworkers also grants manufacturers greater flexibility in meeting market demands by allowing them to diminish or expand the workload as needed. The use of homeworkers gives employers extra leverage in gaining concessions from union workers.

According to Carol Parsons, a former professor of urban and regional planning at the University of Iowa who is studying homework in selected Midwestern industries, the lifting of the ban on homework is just one part of a multifaceted breakdown of labor standards in America.

"Homework," she says, "allows a manufacturer in St. Louis to say to the ILGWU [International Ladies Garment Workers Union], 'OK, you don't want to take a wage cut? Well, we've got 50 women in Iowa who'll work for $2 an hour. It pits the legitimate interests of hard-pressed rural families to earn an income against the interests of urban workers to have the government protect their wages and working conditions. Homework divides groups that have common economic interests."

Because they are spread out across the countryside, rural homeworkers are even harder for labor unions to organize than are homeworkers in cities. A "rural development project" initiated by an Illinois company in Guthrie Center, Iowa, in 1986 is a good example of why industrial homework gives union officials night sweats. Farmers—again, mostly women—in this central Iowa community moonlight assembling front-end suspension components for ITW-Shakeproof, a General Motors subcontractor.

Every week, members from some 60 farm families drive over to the ITW warehouse in Guthrie Center and pick up a "kit" that contains 5,400 individual pieces. The farmers (technically subcontractors themselves) assemble the units out in their tool sheds or in their basements and then return the completed components to the warehouse. According to farmer Connie Benton, she and her husband Ron welcome the opportunity to earn an extra $200 per week.

"Hey," she says, "I'd rather be home instead of having to work someplace over in town all week."

Roger Underwood, the local banker who was instrumental in getting ITW-Shakeproof to locate in Guthrie Center, calls the project the county's first economic development success story after the dismal 1980s.

"Everybody here in town is tickled to death just to have it here," he says.

But Sarah Johnson isn't laughing. Like the Bentons, she and her husband are buying a farm outside of Guthrie Center, and she says that after putting in a full day of farming, assembling auto suspensions is no joke.

"To say that you're tired is putting it mildly," she says.

It's just like an assembly line. It's a full-time job. The worst part is that the work is always there waiting for you. On Sundays. On holidays. It's there when you get up in the morning. And after you're through farming at night. There's just no turning it off. I'm sure they're taking advantage of us. I mean, somebody was getting paid a lot of money to do this in Detroit. But everybody does what they can to survive, and that's just what we're doing. You're just too tired out to protest.

Despite Johnson's doubts, ITW-spokesman Charlie Mitchel thinks that the "subcontracting" system is the way of the future. "More of this kind of thing would be good for these small towns," he says.

You see, we like to have the local community involved. Our particular method is that we don't like to own buildings.

We're not in the real estate business. When we go to a town we like to have somebody there own the building we use as a warehouse. Sometimes a farm widow in a small town will donate the land, and we'll put together a coalition of people and they'll build a building and then we'll lease it back from them. So that way they've got a stake in our business. So that way we're not the bad guys from Chicago who came into town who did this, that and the other thing. It's a very workable system.

Charlotte Reif, formerly a minister in Guthrie Center, has a very different view of that "very workable system," and she expresses that opinion in blunt, angry words.

"It stinks of the piecework my ancestors did sewing in their home after they got off the boat," she says. She became concerned when she noticed how exhausted many of her parishioners were.

"I found out that they were so tired from sitting out there all night in their machine shop doing the most boring of boring line work. It's an injustice. It's slavery. It's a sin, and somebody is going to go to hell for it!"

There are some who maintain that new technologies—primarily computers—will transform the homework industry over the next several decades, leading to a safer, more easily regulated system. But if the past is any guide, new technologies are more likely to reinforce existing patterns of employment than they are to reform them. After all, the introduction of such new technologies as sewing machines, typewriters, and knitting machines did nothing to change the fundamental relationship between homeworker and employer, despite optimistic pronouncements by many that the new machines would do just that. Computers also add new health concerns to the old ones. Video display terminals—especially when used in an unregulated environment—are suspected of causing eyestrain, miscarriages, carpal tunnel syndrome, and a host of other disorders.

Despite these problems, it is likely that homework will continue to spread in rural areas as the newly deregulated industry grows and as unemployment and poverty continue to threaten rural workers. Assurances that these new rural homeworkers will be protected from employer abuse by government watchdogs should be greeted with skepticism. It is extremely doubtful that such an industry can be effectively regulated, as past experience shows.

Before banning homework, the Roosevelt Administration first tried regulating the industry by instituting a certification program that relied on detailed record-keeping. That experiment was judged a failure for several reasons, including difficulties in comparing homework with factory work, a lack of inspectors to check on child-labor violations, and out-and-out forgeries of work records by both manufacturers and workers themselves. Some employers simply premarked employee records with the right number of hours to make it appear as if homeworkers were earning the minimum wage. Fearing that they would be fired for not working fast enough to earn the lawful minimum, many homeworkers lied about the number of hours they worked.[31]

All of these problems exist today. In addition to the widespread wage-and-hour violations found in an investigation of the newly legalized knitted outerwear industry in 1983, an even more egregious pattern of abuse was disclosed by the DOL in the case of Iowa's Bordeaux, Inc. Investigators found minimum-wage violations for *all* of the company's homeworkers, with the back wages due workers totaling almost $730,000. (Bordeaux challenges the validity of the studies on which the department's ruling was made. The company points to an internal DOL assessment that the studies "suffer from a number of weaknesses," including the small number of both seamstresses and patterns checked.[32])

Eileen Boris is not convinced by government promises to keep close tabs on the industry. "There are fewer compliance officers today than there were in 1980—despite a dramatic increase in the number of workplaces that the DOL must monitor. And how can they talk about monitoring the industry when they don't even know how many people are doing homework?" The government's own estimates of the number of people employed in the traditional homework industries range from a low of 8,711 (Bureau of the Census) to a high of 122,000 (Bureau of Labor Statistics).[33]

Despite these problems, supporters claim that homework provides jobs and dollars to a rural economy desperately in need of both. Denying rural women the right to work at home, they say, is a crime. "For many, that [homework] paycheck puts food on the table and may well be the bit of income that keeps the family afloat on the farm," says Iowa Congressman Jim Ross Lightfoot.

These boosters have a point: industrial homework *is* often the only employment open to many rural women. But opening wide the doors to an industry notorious for its low wages and use of child labor is no solution to the tremendous problems facing small-town America. Homework adds no capital, physical plants, roads, or community resources. The support for homework can be explained by the fact that it is much easier to defend a woman's right to choose to be exploited than it is to set about changing the conditions which force her to select that option in the first place.

It is also politically more expedient. To speak out against homework is to risk being branded "antidevelopment" in a time and place that views development (of any kind) as a kind of miracle cure for rural problems. And as with all miracles, religious or secular, the wrath of the believers falls fast and hard on heretics. Probably the most damaging charge that can be leveled against a state legislator from a rural district today is that he or she is somehow not supportive enough of rural development. As a result, many rural advocates feel compelled to modify or limit their opposition on the issue.

"I feel terrible about this," says Peter Brent, a staffperson with the rural advocacy group Prairiefire, talking about the homeworkers employed by Bordeaux.

> These women are hustling their fingers off to sew these things. But if we say anything, we're accused of not caring about the welfare of the state, so we try to keep a low profile. Everybody in this office would like to see an economic rejuvenation in Iowa, and yet it just breaks my heart to watch people have to take these three- or four-dollar-an-hour jobs. I mean that's not a living wage! We deserve better than that.

The latest variety of economic development to receive attention is the rural Enterprise Zone (EZ), a concept that is likely to see increased use in the coming years thanks to support by President George Bush. Modeled after EZs begun in Ireland in the 1960s, the rural American version attempts to lure businesses to rural ghetto areas by offering a variety of incentives including tax breaks for participating companies, relaxed environmental regulations, and exemption from normal minimum-wage laws. Although many call the idea visionary and cutting-edge, the concept is simply the latest reincarnation of the BAWI program of the 1930s.

EZs have had disappointing results in the urban areas where they have been tried. The Government Accounting Office (GAO) studied a Maryland EZ program similar to the federal one advocated by President Bush. The report concluded that "the program did not increase employment growth . . . did not achieve reduction in welfare dependence among workers employed by program participants . . . [and] yielded neither local nor federal program cost offsets."[34] In other words, the program was a failure.

There is no reason to think the idea would fare any better in rural ghettos than it has in urban ones. But if EZs did somehow manage to take root in rural areas, the impact on local communities would very likely be negative rather than positive. As the authors of *The Deindustrialization of America* point out, "Employers who offer low-wage, low-productivity jobs and who follow authoritarian or arbitrary personnel practices, are the most likely to benefit from an enterprise-zone policy. . . . Indeed, there is good reason to expect that the zones could become havens for a revival of old-fashioned sweatshops. . . ."[35]

One Illinois plan amounts to nothing less than an all-out attack on the body of government safeguards built up over the last century through the efforts of unions, environmentalists, and consumer groups. The proposal would establish free-enterprise zones in which virtually anything would go. Employers would not be bound by state minimum-wage laws or zoning or building codes. Property taxes would be abolished, and regulations protecting worker health and safety and the environment would be weakened. Unions in the zones would be effectively emasculated by "right-to-work" laws.[36]

The resulting zones would become the economic and social equivalents of the Galapagos Islands, the Pacific archipelago made famous in the writings of biologist Charles Darwin. Just as unique varieties of flora and fauna developed on the Galapagos, separated from the South American mainland by 650 miles of water, the inhabitants of these rural enterprise zones—cut off from most government regulation and effectively beyond the reach of any ameliorating influences such as unions and media attention—would likewise develop their own society characterized by extreme exploitation and widespread poverty. We can already see the beginnings of just such a society in the spread of rural ghettos.

The many varieties of development practiced in rural America today—from the establishment of enterprise zones to the growth of industrial homework to old-fashioned smokestack chasing—have one thing in common: far from offering solutions to the problems of rural America, they are themselves symptoms of an ill-conceived and antiquated approach to development that does not work for an ever-growing number of rural citizens. Only by taking a wholly different approach—one in which rural communities and their inhabitants are considered to be something more than simply factors of production—will we find solu'   to rural ghettoization.

8

What Future, What Hope?

*What is its [the Midwest's] future? she wondered. A future
of cities and factory smut where now are loping empty
fields? Homes universal and secure? Or placid châteaux
ringed with sullen huts? Youth free to find knowledge and
laughter? Willingness to sift the sanctified lies? . . . The
ancient stale inequalities, or something different in history,
unlike the tedious maturity of other empires? What future
and what hope?*

SINCLAIR LEWIS
in *Main Street*

Fresno County, California

Standing in the dry afternoon heat, vegetable farmer Tom Willey tugs at one end of his bushy moustache and with his other
hand points to a nearby almond orchard where immature nuts
the size of dimes shine hard and green in the California sun.

"The guy who owns that place keeps borrowing money year
after year and never makes it back," says Willey. "He's about 60
and he's built up a hell of an equity, but every year it's just
eroding, eroding, eroding. He's on his knees every night pray-
ing that the damn urban development will come out here and
he'll be able to sell out for $50,000 an acre before he's not
worth anything."

Willey, a short and wiry Los Angeles native in his midthirties,

153

removes his dark green cap with the legend *New Farm,* the
name of a favorite magazine, and wipes his brow. "Man, that is
sick," he pronounces, shaking his head sympathetically at his
neighbor's predicament.

I have come to Fresno County to try to glimpse the future of
rural America. According to a recent government study, if current trends continue, the country's broad-based agricultural system will soon resemble California's corporate-dominated one.
Located in the San Joaquin Valley, Fresno is at the heart of the
state's agribusiness; it is, in fact, the largest agricultural producing county in America. If it were a state, Fresno County's farm
production would rank higher than 25 states', an achievement
which is all the more miraculous when you realize that the region is almost entirely desert.

Despite legislation passed during the 1860s which was designed to settle 50,000 family farms on California public lands,
only 7,000 such farms were created, largely because speculators
managed to buy up huge tracts of land by bribing officials and
filing dummy claims.[1] By 1871 some 122 individuals each
owned stretches of fertile land roughly twice the size of today's
Washington, D.C.—estates 450 times larger than the "family-
sized" farms legislators envisioned.[2]

Much of this land eventually did find its way into the hands
of small farmers, however—enriching speculators in the process. The result in Fresno County was a diverse community of
family farms growing a wide range of fruits and vegetables and
raising livestock, a system similar in many ways to that which
still dominates the Midwest.

Soon after World War I, large corporations and giant incorporated family-owned farms began moving into the area, swallowing up one farm after another, until today agribusiness
giants rule the valley and small family farmers like Tom Willey
and his wife Denesse exist only between the cracks of the enormous agribusiness system.

The Willeys say that it is a combination of smart marketing,
hard work, and luck that allow them to stay in business. They
grow a panoply of trendy vegetables like arugula, French fillet
beans and baby savoy spinach on 35 acres of rented property
just east of the city of Fresno. The only land the couple own is
the one-half acre of fill-dirt beneath their house in town.

"The few family farmers who are surviving now are the ones
who lease their land or who already owned it 15 years ago,"
says Denesse, who is as rotund as her husband is thin but
shares his dark brown eyes and his love for farming. "We joke

about it," she admits, "but the best thing that ever happened to us was when the bank refused to loan us any money."

In the past few months there have been three wholesale auctions in which 20 to 40 local farms were sold at each, and bank officials have predicted a net loss of 25 to 30% of area farms over the next five years. This, in an area already dominated by huge farms.

On a drive through the east side of the valley in the Willey's old Honda car the effects of industrialized farming are not at first noticeable to the untrained eye. All you see is row after row of grape trellises stretching across the flat valley floor, the individual plants appearing brown and stump-like in their early-spring quiescence. The long dry season is just beginning here—it will not rain again until September—and the importance of the irrigation ditches, which crisscross the area carrying murky brown water, is obvious.

But it is what you do not see that tells the story of the San Joaquin Valley. You do not see what we generally think of as farms, with houses and barns, perhaps a few trees, animals, mixed crops, a field here, an orchard there. And you do not see much wildlife. No ground squirrels in the day or raccoons at night. No scraggly cats prowling the roadside ditches. There are certainly no fish in the pesticide-laden irrigation ditches, and there are even few birds in the sky above the fields. Spring in Fresno's grape country—where more pesticides are sprayed than anywhere else in California—is, as Rachel Carson warned, a season of silence.

"This used to be a diversified farming area," says Tom. "Just look at it now. Mile after mile of grapes—and all of one kind. This is a table-grape farm," he says pointing to the fields on the right, where grape stumps about three feet high rest below double-wire trellises. The roads are as straight as the rows of trellises we pass and are named after varieties of grapes.

"We should show him Del Rey," Denesse shouts over the clatter of the engine, half turning toward her husband as we drive past Thompson Street—named after the popular seedless table grape.

"Oh yeah, you've got to see Del Rey," agrees Tom, nodding fiercely. "That is one hell of a place."

Five minutes later we rumble into Del Rey, where the crate is king. What appeared from a distance to be large buildings, perhaps three or four stories tall and the color of slate, turn out to be mammoth stacks of wooden fruit crates that surround and tower over the many metal-walled packing sheds that line Del

Rey's narrow streets. Del Rey is a packinghouse town, where
the produce from the surrounding valley farms is brought to be
processed, boxed, and shipped out.

"This is all there is to this section of Del Rey," says Tom as
we drive through the maze of crates and sheds, the din of the
engine intensified by the walls of bare wood and tin. "Most of
the people who live here are Mexicans who work in the sheds
during the season. It's a dirt-poor town."

We pass through Del Rey's main residential area, a short tract
of tiny houses squeezed between the stacks of fruit crates. We
see a total of three people: two children playing with a tire in
their dusty front yard and a woman who watches them, and us,
from her crumbling front steps.

"Farmers used to pack their own produce," Tom says. "Now,
you've got these corporate packinghouses. Some of them send
their own crews to your farm; *they* pick the crop and put it in
their bins, take it to *their* packinghouse where they pack it under
their label, market it, and ship it out. And they give you whatever the fuck they want to."

Tom is by nature a soft-spoken man, and as he talks his voice
never rises to a level above what is necessary to be heard over
the car's engine, but his anger and bitterness about what has
happened, and continues to happen, to the family farmer in
California is clear.

"See," he continues as Denesse heads the car out of Del Rey
and back out into open farm country, "You don't have the corporations owning all the farms—they don't need to. They just
take over the money-making end of it and let the farmer become a kind of serf. A great system, huh?"

We drive back toward the city of Fresno, past endless
stretches of grape trellises, fields that are punctuated by occa-
sional dark mounds of pomace, the residue of grape stems and
seeds which have been piled into heaps 20 to 30 feet high to
dry and be used as cattle feed. The Sierra Nevada mountains
rise above the valley floor to the east, their snowy peaks turned
purple in the late afternoon sun, a color that is reflected a
shade lighter in the irrigation water beside the road.

As we drive, my thoughts drift back to a conversation I had
earlier that day with a professor at the School of Agricultural
Sciences and Technology, California State University, Fresno
(CSUF).

"The agriculture of the year 2000 and beyond is not going to
be the traditional agriculture we've had for years," he had said
with obvious relish. "Agriculture is really going to be handled

by large corporations, by the chemical industries. DuPont and Union Carbide are going to be doing the farming," he had said—adding, without the least shadow of irony or doubt darkening this vision—"with the appropriate management of the large corporations."

And as we drive through the twilight landscape, the clatter of the car's engine filling the air, it strikes me that both the professor and Tom Willey are predicting a similar future for agriculture; it is in their reaction to that vision that they differ. And for good reason. Since the agriculture school at CSUF receives generous corporate research grants, the professor will be a beneficiary of the brave new world of high-tech superfarms and streamlined corporate control. Tom Willey will be its victim.

As we reach the fringes of Fresno, we pass a section of farmland that is being transformed into suburbia almost before our eyes. With farming unprofitable for the small operator, many farmers are selling out to urban developers who subdivide the land into small tracts which are in turn sold to affluent city dwellers with a yearning for the "country life."

"The Blob," Tom calls the voracious process.

Standing up ahead by the side of the road like a sentry, a brightly painted billboard announces the offering of a new "estate" called Appleseed."Two Acre Parcels Just Minutes From Downtown Fresno!" And about 15 minutes from the wooden-crate canyons of Del Rey, although that information is not to be found on the sign.

"*Jee-zus*," says Tom, pointing to the sole house as yet built on the former farmland. It is a miniature castle, complete with a wooden-shingled turret and topped by a red flag that snaps in the wind. Tom shakes his head in disgust. "Jesus," he says again, this time so softly that I am not at first sure whether it is him I hear or just the wind.

The next day as my plane returns me to the Midwest, I gaze down through spotty clouds at the spring fields below and at the many tiny towns—mere light-colored smudges from that altitude—and try to imagine what those rolling hills that I have known most of my life would look like covered with packing crates and imitation castles.

Conditions in America's rural communities are far worse than is generally recognized. Contrary to national assumptions of rural tranquility, many small towns—even those white-picket-fenced hamlets in our fabled Heartland—today warrant the label "ghetto." No other word so vividly, and yet so accurately, conveys the air of ruin and desolation that now hangs over our rural communities. The word "ghetto" speaks of the rising poverty rates, the chronic unemployment, and the recent spread of low-wage, dead-end jobs. It speaks of the relentless deterioration of health-care systems, schools, roads, buildings, and of the emergence of homelessness, hunger, and poverty. It speaks, too, of the inevitable outmigration of the best and the brightest youths. Above all, the word "ghetto" speaks of the bitter stew of resentment, anger, and despair that simmers silently in those left behind. The hard and ugly truth is not only that we have failed to solve the problems of our urban ghettos, but that we have replicated them in miniature a thousand times across the American countryside.

Some have referred to these depressed regions as "America's Third World," an analogy which sounds like one of those glib but ultimately weak comparisons for which modern journalism is all too well known. After all, the conditions found in the worst of our rural areas are not so terrible as those common throughout much of the Third World. Few, if any, American children actually starve to death in our small towns, and most receive inoculations against the life-threatening diseases that often sweep across other, poorer, continents. True, a growing number of rural children in this country regularly go to bed hungry—even malnourished to the point that their brain development is irrevocably impaired. And it is also true that some rural children whose parents lack medical insurance die or are crippled by diseases that could have been successfully treated had they been diagnosed in time. But bad as the situation is in rural America, it still does not approach the suffering endured by millions living in the Third World.

That admitted, it must also be said that applying the label "Third World" to America's rural areas provides a useful framework for understanding how rural poverty is established and why, once established, the problem seems so intractable, persisting despite huge government farm subsidies and ambitious development projects.

Many of the problems common to the Third World can be traced back to a colonial system in which the region existed primarily to benefit the Metropolis. The mother country was enriched by the relationship, and the colony was impoverished by it. A similar system has helped to produce America's rural ghettos.

In both cases raw commodities are taken from the region (mostly grains in the Midwest, energy sources and agricultural products in the South and West), and value-added products are then sold back to the area's residents. A 1964 government study of Appalachia described this colonial dynamic, though it carefully avoided the word "colony," freighted as it is with political meaning. The report concluded that the root cause of the region's many troubles was "its integration into the economy for a narrow set of purposes: the extraction of low-cost raw materials, power and labor and the provision of a profitable market for consumer goods and services."[3]

Today's disintegration of rural life—the breakup of families, small-town organizations, and whole communities—fits the pattern established by colonial powers throughout the Third World, a process aptly described as "the smashing up of social structures in order to extract the elements of labour from them . . . [and so to dissolve] the body economic into its elements so that each element could fit that part of the system where it is most useful."[4]

Because of the rural decline, manufacturers today don't have to travel halfway around the world to find a cheap and docile work force. Instead of setting up plants in unstable Third World countries—with all the attendant problems of language and cultural differences—manufacturers can turn instead to "smashed-up" rural communities in Iowa, the Dakotas, or Kansas.

Rural sociologist Cornelia Butler Flora has observed striking similarities between rural areas in the United States and Third World societies. She points to a pattern of temporary migration in which the family breadwinner leaves home to work elsewhere for several months at a time. That pattern, which is a normal feature of life in the Third World, is becoming increasingly common throughout rural America. Flora also refers to a "proliferation of survival strategies" in rural areas.

"You used to be able to ask someone in a rural area what they did for a living and get a one- or two-word answer," she says. "Now you get a long list. 'I farm, and I sell some feed, and I do this and that.' It goes on and on."

Where off-farm employment was once a temporary solution for getting through transitory hard times, it is today a fact of life for the vast majority of American farmers. Fewer than one in six farmers had any outside source of income in 1930; by 1977, two out of three farmers depended on off-farm employment for more than half their total income, and better than 90% had some off-farm income.[5] Although agricultural economists commonly refer to this phenomenon as "part-time farming," that term is despised by most farmers, who point out that they continue to farm *full-time* while taking on other work.

Iowan Shelly Loesch is a typical "part-time" farmer. When asked about her family's source of income, Loesch rattles off the following: She and her husband Ted farm 560 acres of corn and beans. They also raise cattle and hogs. Ted is a seed-corn dealer, and he drives a semi part-time for his brother's trucking business. Shelly Loesch, in addition to doing farm chores, raising children (the couple have two children still living at home), cooking, and housekeeping, also makes hundreds of craft items and sells them at fairs around the Midwest. And with a neighbor, she runs a chuck wagon service from early spring to late fall, selling sandwiches and drinks from the back of a converted camper at area farm auctions, household sales, and rodeos. In 1988 alone, the pair worked 26 events.

If we are to reverse the trends that are turning Heartland communities into rural ghettos, forcing residents to adopt Third World survival strategies, we must first confront problems within agriculture. This is because agriculture remains the cornerstone of small-town life and the rural economy—despite current discussion about industrial diversification in rural areas—and a development program that does not give top priority to agriculture is doomed to failure. Once again, the experiences of Third World nations hold important lessons for rural America.

Beginning in the 1950s, rapid industrialization was seen as the key to Third World development. On the advice of international development experts, poor countries throughout Africa, Asia, and Latin America shifted scarce resources from agriculture—their traditional mainstay—and devoted them to ambitious programs of industrialization, building modern factories, massive hydroelectric plants, and oil refineries. With rare unanimity, economists predicted that the strategy would produce worldwide prosperity.

Four decades later, it is clear that rapid industrialization at the expense of agriculture has had a very different result. Not only did the frenzy of "modernization" fail to create widespread affluence, it severely damaged the ability of Third World nations to feed themselves. According to Nobel prize-winning economist

Theodore Schultz, the plan "turned out in country after country to be a disaster."[6] Factories were inefficient and frequently closed due to a scarcity of trained workers and because countries lacked the cash needed to buy spare parts. Oil-rich countries such as Venezuela became dependent on high petroleum prices in the 1970s and then watched their economies fall apart when prices dropped. In a brutal attempt to change a rural society to an urban one overnight, Romania's dictator Nicolae Ceausescu destroyed scores of peasant villages and replaced them with many "agro-industrial" centers, in which industry was never successful and agriculture withered. Countries borrowed millions of dollars to finance industrialization—as well as to purchase imported food when their own agricultural systems foundered—and then found themselves unable to repay their loans when interest rates soared in the seventies.

After decades of pushing rapid industrialization at all costs—a strategy that plunged millions of people into poverty and caused untold environmental damage—economists now say that agriculture must form the basis of healthy development. But just as the experts have realized the catastrophic consequences of neglecting agriculture in the Third World, rural development leaders are poised to make the same mistakes here at home.

Many economists continue to downplay farming's importance to America's rural economy, arguing that an agrarian perspective may have made sense in the 1800s when most rural people (and 40% of all Americans) still lived on farms, but that in a time when farmers comprise less than 10% of the nation's rural population, farming is simply not very important. Yet agriculture's continued importance to rural communities was made clear in a 1985 study of the country's 2,443 nonmetropolitan counties.[7] The USDA study grouped these counties according to their dependence on economic sectors. The nation's 702 farming-dependent counties comprised the single largest category in the study. The importance of agriculture becomes even more obvious if we remove from the mix retirement counties and counties in which more than a third of the land is owned by the federal government. Forty-one percent of the remaining rural counties are dependent on farming, and the bulk of them are found in the nation's Heartland.[8] Rural development and agriculture are as inextricably linked in this country as they are around the world.

Agriculture is itself in a period of transition, one that will have a profound effect on the future of our rural communities. Despite the strong sentiments voiced throughout society for preserving the family farm, the industrialized system of agriculture that already dominates in California is becoming a reality throughout rural America. We are rapidly heading toward a dual-sector system in which middle-sized family farms are all but eliminated, leaving a small number of huge superfarms and numerous small, and economically unimportant, "hobby farms." The total number of farms is predicted to drop from 2.2 million in 1982 to 1.2 million in 2000, with just 50,000 superfarms accounting for 75% of the country's total agricultural production by the turn of the century.[9]

Supporters claim that society benefits from this consolidation because the larger farms, which are highly dependent on capital-intensive technologies, are more efficient producers due to economies of scale. The advocates of industrial farming see themselves as level-headed businessmen who are aligned with the forces of progress, and they characterize defenders of family farming as ignorant and old-fashioned, comparing them to the antitechnology Luddites of early-nineteenth-century England. But the situation is not so simple. The notion that "bigger is better" and the belief that change and progress are synonymous are themselves deeply ingrained American myths. It is just as foolish—and perhaps even more harmful—to uncritically embrace industrialized farming with its grand vision of county-sized superfarms run by chemical companies as it is to cling blindly to some antiquated notion of farming that eschews all modern technologies.

Although economists argue over many aspects of this debate, the majority agree on several key points:[10]

The smallest group of farms *are* often inefficient producers.

Peak efficiency is achieved by medium-sized farmers, those selling around $133,000 in crops annually, employing one or two people, and using up-to-date technologies. The bulk of Midwestern family farms fit in this category. According to a USDA study, "About 80 percent of [medium-sized farms] likely have costs *below* the national average cost of production."[11]

While larger farms are generally no more efficient than medium-sized ones—and they are often significantly *less* so— they provide their owners with greater profits due to the greater volume of crops produced. The largest farms also reap tremendous profits because federal policies—from tax laws to subsidy programs—favor them over the midsized producers.[12] The combination of these factors—a hunger for ever-increasing profits and the existence of market-distorting federal policies— leads to an increasingly inefficient and an increasingly concentrated system of agriculture. It is also a highly unstable system. Because superfarms are almost always superdebtors, they are prone to crises whenever market conditions are unfavorable. And because they are so big, if two or three superfarms in one area do go bankrupt, they can take whole banks with them. Marty Strange, codirector of the Center for Rural Affairs, calls the resulting situation the "Chrysler syndrome."

"We saved Chrysler because we couldn't afford to let it fail," he says. "That's pretty much the situation in American agriculture: There are a handful of superfarms, and we can't afford to let them fail. So we pay more and more to keep them afloat, even though they're terribly inefficient."[13]

The concentration of ownership throughout agriculture (not just in farming, but in all areas of the food producing and processing industry) has important implications for those who live far beyond the fields and small towns of rural America. Conglomeratization in the meatpacking and grain processing industries first hurt farmers by reducing the number of buyers for farm products and so decreasing farmers' profits through the elimination of competition. (The simultaneous mergermania in the farm supply industry also ended up cutting into farmers' profits through increased prices for farm inputs.) But reduced competition in the food industry now also threatens consumers with higher food prices in the checkout lane.

Several segments of this industry are already concentrated to an economically unhealthy degree.[14] Just four firms account for 86% of breakfast cereals sold in America;[15] four companies sell 62% of broiler chickens;[16] three giants sell almost three-quarters of the nation's beef,[17] and the same three—IBP, ConAgra, and Cargill— also control between 30 and 40% of the nation's hog market.[18] The trend is even more pronounced within geographic regions. For example, three companies today control more than two-thirds of the retail food market in the city of Los Angeles.[19]

Thanks in large part to the frenzy of mergers and buyouts during the mid-1980s, agricultural businesses are some of the

largest and most powerful companies on the American scene today. In fact, the three largest private companies in the country in
1987 were agribusinesses—Cargill, Safeway Stores, and Continental Grain. The three had combined sales ($66.2 billion) almost
equivalent to the total receipts of the next ten largest companies
($68.7 billion).[20]

In many respects, Cargill Inc., the largest of the three, resembles a good-sized nation more than a company. Best known for
being the world's largest grain trader, Cargill started out in 1865
with a single grain elevator out on the Iowa prairie. Today, the
company has grown into a vast empire employing 42,000 people
in 46 countries[21] with an annual sales volume ($32.3 billion) equal
to the combined gross national products of Chile and Ecuador.[22]
Cargill is far more than just the nation's number one grain ex-
porter; it is a vertically integrated, horizontally expanding giant,
producing more eggs than any other company in America, rank-
ing second in the nation in beef packing, third in corn milling,
eighth in steel production, and near the top in seed, animal feed,
and fertilizer production.[23]

Ironically, such giants of the "free market" lead to markets that
are anything but free; they create distortions and inefficiencies
generally associated with socialist central planning. Referring to
the effect these giant corporations have on the economy, Pulitzer
prize-winning journalist Lauren Soth warned that "the allocation
of capital, the management of labor, the development of resources
in many leading industries are not determined by the 'unseen
hand' of market forces but by central decision-makers with power
to lead in price-setting and wage policy."[24] When Soviet leader
Mikhail Gorbachev summed up the failures of that country's cen-
trally planned agriculture, he was also voicing the fears of many in
this country over a future dominated by industrialized farming.
"We have," he said, "turned [farmers] from being masters of the
land into mere hirelings."[25]

Industrialized farming, with its dependence on highly toxic
chemicals, monocropping, and a myopic concern for short-term
profits, also poses a serious threat to the environment. In his
classic study of society and the environment, A Sand County Almanac, naturalist Aldo Leopold wrote of the Round River, a "stream
of energy which flows out of the soil into plants, thence into animals, thence back into the soil in a never ending circuit of life."[26]
It has only recently begun to dawn on us that we live along the

banks of another Round River, one that carries death as well as life, destruction as well as regeneration. The garbage we toss so blithely into the river today inexorably flows back to us tomorrow—or if not tomorrow, then next week or next month or next year. The residents of the nation's rural ghettos are unquestionably the first to have their land befouled and their drinking water poisoned by farm chemicals, but they will just as certainly not be the last. On the banks of the Round River, every community is a downstream community. The fact that pesticide residues on domestically raised food cause some 20,000 new cases of cancer in this country every year should bring home to consumers the harsh realities of life on the Round River.[27] As it winds through the global marketplace, the Round River does not recognize international boundaries, and so we must add to the figure above the new cancers caused by pesticides banned for use in the United States but manufactured here, sold for use in other countries, and returned to us on imported foods.[28]

In any event, today's agriculture is simply not sustainable in the long run, no matter how many cancer deaths we can somehow rationalize as "acceptable." Our soil base, though vast, is not inexhaustible, and we are rapidly using it up. The same is true for the supplies of oil and natural gas that are the basis of our synthetic farm chemicals, both pesticides and fertilizers. Eventually, farmers will have to kick their chemical dependency habit and adopt a truly sustainable method of food production, one which depends more on principles of good stewardship and resource conservation than on technological quick fixes. The only question is, What will the toll be in terms of diminished human health and despoiled land and water before we act?

A 1989 landmark study by an arm of the National Academy of Sciences (NAS) provides some hope that a major change in agricultural practices may be in the offing. The study by this prestigious body concluded that farmers would reap economic—not just environmental—benefits if they substituted sound ecological practices such as conservation tillage, biological pest controls, and crop rotation for their present dependence on farm chemicals.[29] Perhaps with increasing public attention focused on the environment the time is ripe for such a shift. As recently as 1981 the United States Department of Agriculture (USDA) compared sustainable-farming advocates to cult followers and largely eliminated research into the field. But eight years later, on the day that the NAS issued its study backing these same "cult-like" practices, one USDA spokesperson enthusiastically endorsed the report, calling it "of unparalleled significance."[30]

And yet it would be foolhardy to read too much into these developments. A system as large and as entrenched as modern agriculture resembles a giant ship: it turns slowly, and then not necessarily in the direction desired by the crew. It is the captains of industry who by and large chart the course for agriculture, through political action committees, through their armies of well-connected lobbyists, through use of the revolving door that connects the agribusiness industry to the USDA, and through all the other advantages which accrue to giant corporations operating in an increasingly concentrated market. Those who now discern the rosy-fingered dawn of a new age of ecological awareness simply because a few magazines ran cover stories about global warming and oil spills should remember that despite the even greater public outrage over pesticides following the publication of Rachel Carson's *Silent Spring* in the 1960s, use of those chemicals continued to grow.

Less obvious, but just as serious as the environmental damage done by industrialized farming, is the threat such a system poses to basic democratic values, for such a highly stratified system of agriculture produces an equally stratified rural society. While the nation's farming-dependent counties as a group have the second highest average per capita income, they also have the next to the lowest median family income.[31] At first glance these figures appear to be contradictory. However, they merely suggest that a very few of the inhabitants of these counties live exceedingly well, while a majority of the population live rather poorly.

The trend toward agricultural industrialization promises to widen these existing gaps. A government study focusing on the most industrialized agricultural counties in California, Arizona, Texas, and Florida (called the CATF region) concluded that

> there is a strong correlation between increased concentration and substandard social and community welfare in this regional set of counties. However this relationship is not strictly linear. As agricultural scale increases from very small to moderate farms, the quality of community life improves. Then, as the scale continues to increase beyond a size that can be worked and managed by a family, the quality of community

life begins to deteriorate. Increasing concentration in this region results in increasing poverty, substandard living and working conditions, and a breakdown of social linkages between the rural communities that provide labor and the farm operators. . . . The most extreme poverty in CATF counties is found in those counties with the most concentrated and productive agriculture. Up to 70 percent of the population of the most highly concentrated counties live in poverty.[32]

Such disparities of wealth—the hallmark of industrialized farming—are inimical to democracy, for in a capitalist society political power is the invisible twin of economic power, and as one concentrates, so does the other.[33] "We can have democracy in this country or we can have great wealth concentrated in the hands of a few," warned Supreme Court Justice Louis Brandeis, "but we can't have both."[34] Nowhere in American society is the truth of this observation more painfully evident than it is in the broken-down homes and along the boarded-up Main Streets of America's rural ghettos which sit amidst sprawling and prosperous superfarms. The concentration of economic-political power in the hands of a few agribusinessmen has resulted in costly federal programs that benefit these superfarms at the expense of both midsized farmers and taxpayers: the largest 15% of American farms captured 70% of direct government farm payments in 1984.[35]

U.S. farm policy, which combines the inefficiencies and market distortions of central planning with the social problems of laissez faire capitalism, is clearly in need of a thorough overhaul. Our current system of paying farmers a subsidy for each bushel of grain they produce is a billion-dollar flop, a bottomless pork barrel that leads to overproduction as farmers try to maximize profits by squeezing as much grain out of their land as possible. It is hard to conceive of a more environmentally disastrous policy, for it encourages farmers to plant on fragile land, deplete water reserves, and pour on farm chemicals to boost production. What is needed is a federal farm policy that ensures a stable supply of nutritious and safe food at reasonable prices, protects the environment, and promotes the well-being of the efficient medium-sized farms that are the lifeblood of rural communities. Rather than pegging government subsidies to total bushels produced, some have suggested paying farmers only for a certain number of bushels of grain—the "quota" for an individual farmer to be calculated on the basis of a variety of factors including farm size and environmental conditions.[36]

Far more needs to be done to help agriculture achieve its potential as an efficient, sustainable, and equitable industry. Research at land grant institutions should reflect the interests and needs of small and medium-sized farmers, consumers, and taxpayers—not giant agribusinesses as is often the case. Tax laws that encourage overcapitalization in agriculture by investors—not farmers—need to be changed. Minimum-wage laws and occupational safety and health regulations which currently do not apply to farm workers should be extended to them.

But reforming agriculture policy so that efficient family-sized farms are given a chance to prosper is only a starting point in crafting a strategy for rural development. Once that solid foundation has been laid, a more general program of rural development must be undertaken.

From the postbellum South to the farm-crisis Midwest, the primary mode of rural development in America has always been job recruitment—at all costs. Born in a climate of decline and desperation, this form of economic development sets communities, states, and whole regions against each other in a destructive competition to attract jobs by giving up more and more of themselves economically, environmentally, and socially. Development based on the mistaken notion that job growth leads inevitably to community prosperity simply reinforces the cycle of decline it is designed to end. This form of development—the product of an outmoded industrial-age mentality—cannot adequately address the challenges or take advantage of the opportunities facing rural communities in a postindustrial age.[37] Only *community* development, which goes beyond the blind allegiance to job recruitment found in purely *economic* development, can.

Community development takes into account the pay scale offered by new jobs, as well as the environmental and social impacts of specific development projects (including the differing impacts on each segment of the population: male and female, elderly and young, wealthy and poor, majority and minority group). This kind of development asks many questions: What demands will a new industry place on sewers, roads, schools, social services, police? What benefits will a new project provide, in terms of high-quality jobs, recreational possibilities, civic participation?

In short, community development takes into account all those factors which together determine the *quality of life* enjoyed by residents.

Such a policy makes good business sense. Americans typically move in pursuit of "the good life"—not in pursuit of jobs alone—[38]and the rural Heartland still affords residents low crime rates, an unhurried pace of life, and an emphasis on traditional values of family, hard work, entrepreneurialism, education, and civic participation. Many of these assets are now eroding, but they could be maintained and strengthened if given the same attention currently given job recruitment. Community development should not be seen as a call to overlook or deny the tremendous problems of rural communities. It is, instead, a reminder of the equally tremendous potential these areas have for fulfilling the American dream.

Deplorable as they are, conditions in rural communities today would be far worse if the country had not endorsed a policy of supporting rural development long ago. Since the first federal studies on rural road construction in 1892, government planners have seen the necessity of supporting a viable countryside. A hundred years later, we must renew and expand on that commitment or face a country seriously, and perhaps permanently, divided.

It has been pointed out that "America cannot survive half rich, half poor; half suburb, half slum,"[39] and that observation is just as sound whether the slum is found in a crowded inner-city neighborhood or in some backroad town languishing on the prairie. But we must take action soon if we are to avoid establishing a permanent underclass in the American Heartland. The persistence of poverty in our inner cities in spite of several decades of federal programs suggests that once the wheels of decline have been set in motion, they are difficult to reverse.

Rural America is not an island apart from the rest of the country, however, and while a program of rural renewal is essential if we are to counter the effects of recent hard times, ultimately the future of small towns hinges less on the resolution of narrowly defined "rural" issues and more on our ability as a nation to tackle the larger conundrums of our political economy. Putting teeth in our antitrust laws, strengthening the labor union movement, making our tax code more equitable, instituting election reforms so that voters, not political action committee dollars, determine who will serve in Washington, D.C.—accomplishing any one of these would do more to help the nation's 10 million rural poor than would throwing more money at farmers or scattering a hundred new "high-tech" assembly plants across the land. To heal our broken Heartland will take something a thousand times harder to come by and infinitely more precious than economic growth. For in the final analysis what rural America needs most is not more jobs or more money, but more democracy—in the form of a citizenry willing and able to participate fully in the development and sustenance of their communities.

It is fashionable to dismiss Thomas Jefferson's agrarian society as an outdated utopia which was, in any case, restricted to white men. But while there is much to criticize both in Jefferson's original vision and in how sparingly it was actually implemented, the democratic principle central to Jefferson's ideal—the commitment to community assured by the yeoman farmer—remains our passport to the future. The challenge is to adapt that eighteenth-century conception of society to fit the realities of the twenty-first century. If we can meet that challenge, then the golden age of rural America will lie not in our past—as our myths have it—but in our future.

Coda: 1996

When loss of community commitment becomes the rule rather than the exception, there is little hope for [the] future. While [the average small town] is not at the point of such despair, if nothing is done to reverse ongoing trends, there is little to stand in the way of such a fate.

VERNON RYAN, ANDY TERRY, AND DANYAL WOEBKE
In *Sigma: A Profile of Iowa's Rural Communities*, June 1995

Imagine you're seeing them for the first time, perhaps as they're buying groceries at the Food Lion out on Highway 74 in Hamlet, North Carolina. Their names are Josephine Barrington and Rosie Chambers. What do you notice as you follow them down the aisles? Probably what strikes you first is how different they are. There is the matter of age. In 1991 Josephine was 63 years old, a grandmother, with a lifetime of hard work behind her. Rosie was barely out of her teenage years. She was also more ambitious than the older woman. Rosie was struggling her way up the economic ladder, working all day, studying for a few hours in the afternoon, and then driving her blue Ford Fiesta over to class at Richmond County Community College.

But the women also had a lot in common, and that would be evident, too. Both were black. Each was a faithful churchgoer and sang in the choir, Josephine at the Rock of Faith Temple Miracles Church of God in Christ. Rosie had a rich full voice and sang solos at the Powerhouse Church of God. And both women worked for poverty wages at the Imperial Food Products poultry plant. Where else *would* they work? Hamlet was once a prosperous railroad town, but it had been on a downward spiral for decades. There were few jobs left in the area, especially for unskilled workers like Josephine and Rosie.

The two women had something else in common: they each hated their job at Imperial.

It wasn't just the poor pay. The women complained bitterly to friends about the terrible conditions inside the factory. They told horror stories of power lines lying in pools of water. They talked about the unbearably hot processing room—called "the hole"—where pieces of breaded chicken were fried in 400-degree oil, which sloshed over the edges of the cooking vat, burning their legs. They grumbled about the refrigerated packing room, where women worked with cold-numbed fingers, cutting themselves on razor-sharp knives and not realizing what they had done until they left the room and their hands thawed out and they began to bleed.

Even worse than the physical hazards at the plant were the assaults on workers' dignity. The women were allowed just two bathroom breaks a day. Use the bathroom at an unscheduled time, or take too long in the stall, and you were written down for an "occurrence." Five occurrences and you were fired. Grown women had been known to wet themselves rather than risk losing their job.

Neither Josephine nor Rosie had complained to management about the working conditions at Imperial. They had heard the supervisor's response to those who protested: "There's the door. Use it."

Both women were in the factory working the first shift on the morning of September 3, 1991, the day after Labor Day. At 8:20 a.m. a hydraulic line gave way, spewing motor oil onto the open flames of the large cooking vat. There was a flash of light as the oil ignited, and then the room quickly filled with smoke. A hot black cloud raced through the factory.

Josephine and her 37-year-old son, Fred, also an Imperial employee, tried to escape through a door marked "EXIT," but it had been padlocked from the outside. An Imperial worker later said it had been locked to prevent employees from smuggling out pieces of chicken. When the door wouldn't budge, mother and son joined ten other workers inside a walk-in meat cooler. It isn't known whether they blundered into the cooler, mistaking the door for another exit, or simply believed the metal walls would protect them from the flames and smoke. When fire fighters entered the gutted building, they found all 12 employees dead on the cooler floor. Josephine Barrington's body was cradled in her son's arms. Rosie Chambers's body was found crumpled against a second locked door, less than two inches separating her from fresh air and from life.

Twenty-five people died at the plant that day. The death toll could have been much higher if a group of workers hadn't managed to kick down a third illegally locked door. The media called it the worst industrial accident in North Carolina's history and the worst in the country in years. But calling what happened to Josephine, Rosie, and 23 others an "accident" distorts language. The deaths were a result of economic development based on desperation. Imperial (has there been a more aptly named company?), headquartered in Pennsylvania, was just one of dozens of corporate carpetbaggers drawn to rural America in the past decades, attracted by the promise of low taxes, weak unions, desperate workers, and government agencies willing to overlook violations of safety and health laws in the name of jobs. Not a single government safety inspector had entered the plant in the 11 years Imperial had been operating in Hamlet.

Josephine Barrington and Rosie Chambers died two and a half years after I submitted the manuscript for *Broken Heartland.* In that time, and since, conditions in rural America have changed very little, overall. An optimist might characterize the period from 1990 to 1996 as a plateau in which rural problems, on balance, didn't get much worse.

There have been brights spots amid the general gloom. The financial crunch dubbed the "farm crisis" lessened for many farmers—at least for those who survived the 1980s. For many others, mostly small and medium-sized farmers, the future is still in doubt. These farmers managed to hang on through the worst years with the hope that the credit squeeze of the '80s would give way to a more "normal" period in the '90s in which farming would once again become profitable for mid-sized producers. But agriculture did not return to its glory days. Mid-sized, diversified operations worked by families are still on the way out, just as they had been before the 1980s.

On the plus side, "sustainable agriculture," which was once regarded as a slightly kooky crusade, has entered the American mainstream. The average farmer today applies less pesticide per acre than in the past. Many rely on a variety of nontoxic techniques to keep weeds and bugs in check. Soil loss has been significantly reduced by farmers practicing good land stewardship. These achievements should not be slighted. But our larger political economy still favors giant corporate interests at the expense of family farms and rural people in general. The resulting hybrid of low-input agricultural techniques and transnationals freed from government regulations—and from the strictures of national borders themselves—may herald a new epoch: the age of Sustainable Exploitation.

Again on the positive side of the ledger, there has been a boom in rural jobs. Employment outside of metro areas continued to decline as the decade opened, but then suddenly the situation turned around. Rural jobs rapidly became more abundant. Since 1990, rural jobs have grown at a rate double that of the 1980s. Between 1993 and 1994, rural employment expanded by 2.8%—the biggest jump in nearly two decades—far outpacing the metro job growth rate.[1] Like ancient oracles divining the future in charred sheep bones, some experts see a bright future for rural America in these statistics. Once again, it is "morning in [rural] America." In May of 1996 *Governing* magazine ran a cover story bearing the optimistic title, "Return to Main Street," declaring that "the small-town comeback is no illusion." Indeed, some midwestern towns have shown signs of new life (although almost all of them are medium-sized by midwestern standards, not truly small towns).[2]

But job growth and the resurrection of some towns haven't translated into widespread rural prosperity. This becomes clear in the *Governing* article when the author admits that "many of the tiniest [towns], a few thousand people or less, have all but gone under . . . and have no realistic chance of scrambling back."[3]

Why hasn't prosperity followed job growth? There are several reasons for this apparent contradiction. First, while these employment statistics appear impressive on paper, the view on the ground is much different. There were so few new jobs in rural areas by the late 1980s that any increase looks like a boom. Even more important, many of the jobs moving into rural areas pay low wages and offer few if any benefits. A rural worker today is even less likely to earn a living wage than in 1987. As a result, even with both adult household heads working, often commuting 50 miles or more to their jobs, median rural household income actually fell 3.2% between 1989 and 1993.[4]

Too, the traditional multiplier effect–in which each new job creates other new jobs–is not operating in these new industries. These mostly small manufacturing plants purchase their inputs from distant sources and truck them in, rather than trading locally. Their only local purchases are of water and electricity–not the stuff sustainable development is built on.

And don't expect the job boom–such as it is–to continue, warns David Swenson, an economist at Iowa State University. These new jobs do not come with a promise of security.

"For a lot of rural plants, it doesn't matter to them whether they're in northern Iowa, northern Missouri, or northern Mexico," according to Swenson. "One reason our rural places became attractive is we had a surplus of relatively skilled, relatively intelligent labor after the farm crisis." Now, says Swenson, that supply has been used up. "The pattern of growth–where it's occurred–has established itself, and the likelihood of continued growth, not just in Iowa, but nationally, is going to go down."[5] Some manufacturers have already left. Others will leave the region and set up shop on the U.S.-Mexico border as soon as the tax concessions they wrung from their desperate host communities run out. Like the capitally intensive agriculture it replaced, a significant share of the current rural-jobs boom rests on the thinnest of crusts. Below, just out of sight, sits the same sinkhole that swallowed communities whole in the 1980s. The best that can be said for this jobs boom is that it has, at least for the moment, stabilized some communities–generally those adjacent to larger population and manufacturing centers.

Despite periodic stories of rural revitalization and renewal, the persistence of poverty in rural America is like an arrow shot into the multicolored balloons of optimism. Rural poverty rates did decline from their highest farm-crisis levels (18.3% in 1983), down to 15.7% in 1989. But despite the end of the farm crisis, poverty has grown in rural America, not diminished. By 1993 the rural poverty rate had shot back up to 17.3%.[6] Again, there are several reasons for this. First, agriculture is not a classless industry. It's nonsense to speak, as pundits and politicians often do, of a particular farm bill as being "good for farmers" or "bad for farmers." When larger farms are doing quite well, it is often at the expense of smaller farms. Second, there is much more to rural America than just the agricultural sector. If all farms became profitable overnight, large parts of rural America would remain mired in poverty. Even in the Heartland other industries are increasingly important to rural health.

Outside of the Midwest, the rural economy depends on a variety of industries: mining and logging, manufacturing and service, most of which are having their own problems.

Since *Broken Heartland*'s appearance in 1990, one of the most contentious issues in the rural Heartland has been the consolidation of the hog industry. What happened to poultry and to cattle in the previous decade is now occurring to hogs: family-sized operations are being pushed out of business by giant hog lots. Former Secretary of Agriculture Earl Butz may be out of the picture, but the "get big or get out" mentality he made infamous remains the driving force of our age. And hog lots are getting huge: raising hogs is no longer a matter of running a few dozen sows on an idled cornfield and producing a few hundred hogs a year. Today, hogs are raised by the tens of thousands in neat rows of windowless metal buildings the size of football fields. A typical state-of-the-art hog lot replaces the work of dozens of family farms. And when one of these operations moves into a county, the results can be devastating.

In 1994 I received a phone call that suggested just how disrupting the arrival of one of these giant facilities can be for local people. It wasn't simply that the caller was irate about the huge hog factory that had located next door. It was the caller's identity that set off alarm bells in my head. On the other end of the line was my former mother-in-law, the mother of the woman I had divorced under not-completely-friendly terms close to a decade and a half earlier. "Osha? This is Dorla Hill," she said. "You've got to write something about what's happening here with these hog lots. It's terrible." I was shocked to hear her voice. Our relationship had always been a bit strained—even before the divorce. For her to call asking for my help now meant that the situation must have been terrible, indeed.

The Hills are third-generation farmers, in their mid-sixties and about as conservative as they come. Dorla's husband, Myron, has farmed the table-flat land of north-central Iowa, outside the town of Clarion, since he was a teenager. I knew him as a gentle, quiet man, wiry, plainspoken and hard working, quick with his shy smile and not one to hold a grudge. "Live and let live" could have been his motto—if he had gone in for something so high-flown as a motto. He had had some hard blows in life, particularly the death of his only son, Larry, a boy as gentle and unaffected as his father, lost to a heart defect in high school some twenty years before. But Myron never became embittered. Until now.

"As far as I'm concerned, they're all rotten down there in Des Moines," he says in 1996, citing the legislature's failure to rein in the giant hog lots springing up all over Wright County. "They call it economic development, but they don't realize the consequences. And of course, as long as they don't smell it, it doesn't bother them."

The smell is the problem critics of these giant hog lots cite most often.

"It's just unbelievable the smell that comes off of there," Myron tells me over the phone, his voice sometimes angry, but mostly just subdued. "Remember, I've raised hogs. I know what hog manure smells like. This is different. It's unbelievable."

I remember very well the hogs on the Hills' farm, back from my first visit to the farm in 1973 when I was their daughter's long-haired and scraggly bearded boyfriend. I had never been on a working farm before and was fascinated by the hogs roaming around a pen beside the barn, grunting merrily as they rooted in the pungent layers of dirt, manure, and straw. Myron ("Mr. Hill," I called him then) was treating me to a quick tour of the farm. He was clearly proud of the spread he had put together. That pride showed through—even to a city-bred college kid who wore bib overalls because they were "cool." As we stood admiring the hogs, Myron glanced over to catch my attention. Then he made a show of inhaling the odor that rose from the pen in a small stinky cloud. "Ah," he said, smiling and throwing in a wink, "the smell of money." It was the first time I had heard that common rural witticism. I was so green then that I thought Myron had just coined it, and I laughed uproariously.

In an average year, the Hills farmed some 500 acres of corn and beans and raised between 300 and 400 hogs a year. A couple of times they raised as many as 700 hogs. The sows gave birth to their litters in the distinctive A-shaped farrowing houses that used to dot the countryside. The houses, just large enough for a sow and her offspring, sat in one field one year and were hauled to another field the next.

"Back then, I considered manure a good thing," recalls Myron. "I could raise hogs and then next year I could raise good corn out there. Now they're just trying to get rid of it."

Myron was referring to the activities of his new neighbors: a factory "farm" which rolls out chickens and hogs in numbers that old-timers find incomprehensible. First came the chickens, mostly laying hens, their huge metal buildings erected on the former farmstead of the Hills' close friends. Other giant hen houses followed. Today, says Myron, there are 6.4 million chickens within three miles of the Hills' farm. That was disturbing enough, but the trouble really started when the hogs came in.

Built beside the chicken houses, just a mile and a half from the Hills' property, stand ten mammoth hog houses. Together, they house 26,000 hogs. These are "finishing units" where hogs stay just long enough to fatten up for slaughter. That means a total of 78,000 hogs pass through this one piece of land each year. Each hog produces up to 15 pounds of manure a day, for a total of 71,000 tons of manure each year. What was a good thing—free fertilizer—on a small scale for a family farmer like Myron Hill becomes a problem for the owners of hog lots who must find a way to dispose of it. So far, their best (i.e., cheapest) answer has been to make huge lagoons of the stuff, where, in theory, it decomposes into a harmless and nutrient-rich slurry which is then sprayed on fields. For neighbors living downwind from these lagoons, "the smell of money" has become a withering stench that drives them indoors on warm days, and, more frequently all the time, off the land entirely.

"It has ruined our whole environment," says Dorla. "We cannot live here. We cannot continue to live here. We cannot stand the odor. We've become angry and mean. It's made us that way. Our thinking has totally changed because of this whole thing."

But the noxious smell is only the most blatant problem caused by these open cesspools. In North Carolina, where giant hog lots first made their appearance, scientists have found other problems. Although the lagoons are supposed to be self-sealing, preventing contamination of local water supplies, one study determined that half of them were leaking into the groundwater.[7]

Around Clarion, the soil is far less sandy than that in North Carolina. Defenders of hog lots in Iowa claim that this difference prevents the contamination of local groundwater. But there are other problems. Much of the land around the Hills' farm had drainage wells drilled in the fields during the early part of the century. These wells allowed standing surface water and water from soggy topsoils to drain down deep into the earth. Today, they also provide a direct route for liquid hog manure to contaminate the area's aquifer. Once the manure is spread on fields, the dangers posed by the process do not end. Soil can absorb only so much of this nutrient-rich matter. What in small doses is fertilizer can in large doses actually kill the soil. When the earth is saturated with manure, rain washes the contaminants into the area's creeks and rivers, polluting them.

And these are just the hazards posed by the hog lots on a daily basis, when things are operating pretty much as they're supposed to. What happens when something goes really wrong, when there is a major accident?

That question was answered in North Carolina in late June of 1995, when the dike surrounding a 12-foot-deep lagoon of hog feces and urine broke, allowing the contents to flow across nearby fields before spilling into the New River. Much of the 25 million gallons of manure from the lagoon entered the river, killing thousands of fish and contaminating water for miles downstream. Worst of all, the site of the accident wasn't some ancient lagoon built before the risks involved were understood. It was a state-of-the-art, nearly new holding pond, designed under what had been termed "stringent" new state regulations.

Despite these risks to the environment, giant hog lots, like the huge poultry houses and cattle feed lots that preceded them, are the way of the future in America's Heartland. Opponents certainly can't match the political influence enjoyed by backers of hog lots, and regulatory efforts in state legislatures have gone nowhere. Nor can opponents compete with the market clout that gigantism provides. Hog-slaughtering facilities used to accept shipments of a few dozen hogs from independent farmers, often at local stations throughout the state. But when packing plants have shipments of a thousand hogs or more arriving regularly from hog lots, there is little economic incentive to maintain a "farmer-friendly" system that can accommodate irregular and small shipments. During the farm crisis years of 1982 to 1987 some 21% of Iowa's hog farmers went out of business. But even after the end of the farm crisis, pressure from these new giant factories caused more hog farmers to call it quits: between 1987 and 1992 nearly 12% of Iowa's remaining hog farmers gave up. Wright County, where the Hills farm, has seen a similar decline, losing 35% of its hog farmers between 1982 and 1987 and another 10% over the next five years.[8] But the explosive growth of giant hog lots has occurred since this last (1992) census, and the next one in 1997 will likely find an even more dramatic loss of hog farms.

When a giant hog farm moves in, the entire surrounding community is thrown into an uproar—those living upwind, too. Such a traumatic event forces people and institutions to take sides, causing painful schisms that can take decades to heal, if they ever do. The Hills joined a group fighting hog lot consolidation. They marched with other farmers in Missouri and protested at the state capitol building in Des Moines. Their activities attracted attention in the formerly placid town of Clarion.

"We've been called rebels and all sorts of things because we're fighting this," says Dorla. "But when your quality of life, where you've lived for all these years and raised your families, when that is threatened, it makes you angry."

Some of that anger is directed at those who either support the giant hog lots or who, as Dorla and Myron see it, stand on the sidelines and allow their neighbors to go under.

"We have not had the support from this area that we feel we should have had," says Dorla. "Our churches are not supporting us. Our businessmen are not supporting us. They don't care. Of course, a lot of them are making money off of this."

And Myron has mixed feelings about other farmers who would sell their land to a hog lot developer. "People got the right to sell ground, I guess," he says. "Of course, money is the most important thing anymore. Farmers aren't neighbors like they used to be when I grew up. It's one neighbor against the other. Now, we're called revolutionists because we're 'stopping progress.' But this isn't progress."

This is all very tough talk from a couple whose idea of "radical" just a few years ago would have been to skip church and lie around the house on a Sunday morning. But things are changing rapidly. Myron looks around, surveying the most fertile land in the world, and sees a future devoid of family farms.

"There won't be any little farmers," he says in a voice as flat as his fields. "I know what's happened to gas stations and grocery stores. So I suppose it's due to come to farming."

That view is apparently shared by Iowa's state officials. The Department of Economic Development recently granted a company a $230,000 loan to build equipment for giant hog lots. The manufacturer was hailed for its part in the "transformation from small family farms to more efficient, large-scale confinement operations."[9]

Toward the end of our conversation I ask Myron what will become of his farm. The act of handing down the family farm to a son or son-in-law is still a defining moment here in the patriarchal Heartland, a ritual straight out of the Old Testament. Handing down the family farm is the culmination of a man's work, when his life is judged to be a success or a failure. For what farmers raise is not simply corn and beans, hogs and cattle, but a *farmstead*, to be improved upon and bequeathed to the next generation in a mystical melding of family and land that is supposed to last forever. The process represents, over time, what Willa Cather's prairie represented in space: the blissful opportunity "to be dissolved into something complete and great."

Before our marriage soured, Myron's daughter and I discussed the possibility of moving to Wright County and taking over the family farm operation. So my question, asked now, nearly fifteen years after I ceased to be Myron's son-in-law, comes with a certain history between us.

What will happen to his farm?

"I don't know," he says, not skipping a beat. "I don't have any son in laws who want it."

Is there an accusation sown into those words? I can't tell. But even as I am considering that possibility, Myron is continuing along a more pragmatic line.

"Maybe it's a good thing," he says. "What would I be leaving? Some big farmer'll probably come in here and take all the fences down and it'll be . . ." He trails off.

"I'd hate to see that happen," I tell him. I realize later that my own statement is a complex message: part sympathy and part apology—with a dollop of encouragement thrown in.

But Myron rejects it all. He hasn't an ounce of sentimentality left in his voice when he says: "Well, it's going to."

The bitterness so evident in Myron and Dorla Hill's voices is reminiscent of other voices from rural America in the 1980s: a priest explaining how economic forces were whipsawing his parishioners, a small-town merchant forced to close his doors after a Wal-Mart was thrown up on the edge of town, a nurse who watched people die because, lacking health insurance, they delayed seeking treatment of a serious illness. And, while I know that the Hills have too much good sense to go this route, the sense of betrayal in their voices is an echo of the call of the radical right that I heard throughout the region in the 1980s.

Those voices are heard again today. When the farm crisis was declared over and the country moved on to other issues, rural people's anger over the break-up of their communities and the chaos of their lives didn't just go away. It festered. And then it exploded.

It took the bombing in Oklahoma City—with its 169 deaths—for the nation to discover that violence-prone hate groups were active in rural America. The militia movement (and the larger Patriot movement of which militias are one part) is the latest manifestation of the loosely affiliated rural network that grew during hard times in the 1980s. No one who tracked these groups was surprised by what happened in Oklahoma City. On the contrary, organizations such as the Center for Democratic Renewal in Atlanta, Georgia, and the Southern Poverty Law Center in Montgomery, Alabama, had been trying for years to get Americans to pay attention to this alarming trend. Before Oklahoma City, however, no one wanted to hear about what was considered a small number of crazy-but-harmless right-wingers out in the middle of nowhere raving about "federal posts" and "race mingling." Let the kooks rant, was the prevailing attitude.

The dangers of ignoring this movement are now heartbreakingly evident. Images of carnage and destruction in Oklahoma City are burned into the American psyche. The fire fighter clutching a child's lifeless body. The blown-out side of the Murrah Federal Building, dripping chunks of concrete and twisted metal. The tragedy in Oklahoma City got America's attention at last. Law enforcement now takes these far-right groups more seriously. The press reports regularly on these groups. Some private foundation money has flowed to organizations working to expose purveyors of hate. In its recent landmark report, *False Patriots*, the Klanwatch Project of the Southern Poverty Law Center spelled out in harrowing detail who these self-described "patriots" are and the nature of the threat they pose:

> The Patriot movement is not easily defined. It has no single national organization or leadership but consists of previously unrelated groups and individuals who have found a common cause in their deep distrust of the government and their eagerness to fight back. They are convinced that the American people are being systematically oppressed by an illegal, totalitarian government that is intent on disarming all citizens and creating one world government. They believe that the time for traditional political reform is over, that their freedom will only be secured by resistance to the nation's laws and attacks against its institutions.[10]

Progress has been made in understanding and spotlighting the far right. What is still lacking is a full-scale attack on the terrible economic and social conditions that provide these groups with a sympathetic audience. The Patriot movement may preach racial hatred, a fanatic's loathing of all gun control, and a paranoid worldview in which U.N. troops zoom around the country at night in black helicopters. But it is the nightmarish *realities* of contemporary rural life that give far-right propaganda its credibility and appeal. After all, the fact that corporate predators such as Imperial are allowed to operate without government oversight is just as sinister as some of the half-baked conspiracy theories about a New World Order. And what could be more pernicious than a giant hog lot, with its lake-sized cesspool, plopped down amid family farms?

What I wrote in 1990 is, sadly, still true in 1996: "As the rural economy continues its slide, the beachhead established by the far right during the farm crisis of the 1980s will, in all likelihood, continue to grow in the 1990s. If genuine alternatives are not provided, a significant number of rural ghetto residents—bitter, desperate, and increasingly cut off from the nation's cities—are sure to seek their salvation in the politics of hate."

And those alternatives do not appear forthcoming in 1996. Indeed, what passes for political discourse today revolves around the question of whether government should be virtually eliminated or merely crippled. Even the lip service formerly paid to saving family farms and preserving rural communities has been replaced by a naive faith in the ability of the "invisible hand" of the market to order society by itself. Never mind that Adam Smith himself warned that market concentration is a threat to civil society. To all questions—economic, political, and social the free marketeers respond: "The marketplace will provide." And although polls show that millions of Americans harbor nagging doubts about the future, all they are offered are these laissez-fairey tales and the vague promises of politicians to uphold "traditional values."

Ironically, it is the most fundamental American value—democracy itself—that is jeopardized by the changes sweeping through the Heartland. It is not the destruction of the family farm system per se, or of individual rural communities themselves, that is most troubling for the American soul. After all, for two centuries American towns have blossomed during periods of economic boom and then succumbed to the inevitable bust that followed. The question, then, is not should we preserve family farms or rural communities but rather *what is replacing them?* The answer thus far is chilling: A countryside dominated by superfarms, corporate hog lots, and factory towns where people such as Josephine Barrington and Rosie Chambers labor for poverty wages in unsafe and unhealthy conditions, a region in which the ties of community are abraded by economic and social chaos. Small towns and the family farms surrounding them formed the cradle of American democracy. When these institutions are gone, where will democracy flourish?

A lesson from field ecology has the last word here, for democracy is a living thing—destroy its habitat and it too will perish.

Notes

Chapter 1
Decline and Denial

1. There are several good sources for reading the accounts of settlers, including Lingeman (1980), Madison (1982), Webb (1931), Smith (1950), and Zinn (1980).
2. Sage (1974:275).
3. Stories about Mechanicsville's history came from a variety of sources including Iowa county histories and interviews with residents.
4. Nye (1961:4).
5. Iowa State University (1981:6).
6. Erikson (1976:193).
7. This difference between rural society (Gemeinschaft) and urban society (Gesellschaft) is discussed in Erikson (1976).
8. Iowa State University (1981).
9. Porter (1989).
10. Ross and Danzinger (1987:2).
11. U.S. Congress. Joint Economic Committee. Subcommittee on Agriculture and Transportation (1986:288).

Chapter 2
Roots of the Farm Crisis

1. Solkoff (1985:43).
2. Moberg (1988:205).
3. Morgan (1979:10).
4. *Des Moines Register* (1986h).

5. U.S. Congress. Joint Economic Committee (1986:169).

6. Staten (1987:39).

7. Moberg (1988:205).

8. U.S. Congress. Senate (1986:6).

9. U.S. Department of Agriculture. Economic Research Service (1983).

10. Crevecoeur (1957:36).

11. Jefferson (1964:157).

12. Jefferson (1964:158).

13. Zinn (1980:95).

14. Benson (1971:108).

15. Benson (1971:123).

16. Many of our early leaders were themselves speculators who made their fortunes in land. By 1796, George Washington himself had acquired 32,373 acres of land—a holding equal in size to two Manhattan Islands. The real-life Daniel Boone was quite different from the Hollywood figure who was portrayed as a romantic rustic looking for adventure and simple living in the wilderness. Boone was actually looking for tracts of fertile land as he meandered through the Kentucky territory. The frontiersman was working for Judge Richard Henderson, a wealthy land speculator from North Carolina who, with Boone's help, amassed a holding of 400,000 acres. The holdings of these men were modest, however, when compared to those of speculating companies such as the Illinois Company and the Indiana Company, which owned 1.2 million acres and 1.8 million acres respectively. While these huge estates were not as oppressive as their feudal antecedents (despite Jefferson's comparisons), neither were they the stuff of which a nation of yeoman farmers is made. See Opie (1987).

17. Geisler and Popper (1984:8).

18. Opie (1987:70–84).

19. Geisler and Popper (1984:13).

20. Zinn (1980:276).

21. Goodwyn (1978:20–26).

22. Goodwyn (1978:23).

23. Goodwyn (1978:297).

24. Goodwyn (1978:297).

25. U.S. Congress. House (1937).

26. Geisler (1987).

27. Myrdal (1944:251–278). See also Abrams (1939:60–77), Piven and Cloward (1972:200–205), and Zinn (1980:388).

28. Wessel (1983:21).

29. Benson (1956:4).
30. The red-taint tactic used so effectively by Benson to silence opponents of his free-market farm policies became standard use for the Department of Agriculture in defense of its policies. In September 1988, Farmers Home Administration (FmHA) head Vance Clark told members of the Downtown Rotary in Janesville, Wisconsin, that farm policies favored by liberals in Congress "will lead to socialism unless we do something about it." *Des Moines Register* (1988t).
31. Morgan (1979:98).
32. Ford (1973:92).
33. Belden (1986).
34. ConAgra (1985:1).
35. Quoted in Krebs (1988:20).
36. *Des Moines Register* (1987g).
37. *Des Moines Register* (1987g).
38. *Des Moines Register* (1987g).
39. *Des Moines Register* (1987f).
40. In *An Economic Interpretation of the Constitution of the United States* (1986), Charles Beard points out that not one participant of the 1787 Constitutional Convention in Philadelphia represented the class Jefferson called the yeoman farmer.
41. For a detailed explanation of the deal, see Solkoff (1985:46–56).
42. Solkoff (1985:55).
43. Solkoff (1985).
44. Solkoff (1985:49).
45. Kaplan (1987:13).
46. CAN won its case in California state court in 1987, but the university won on appeal in 1989.
47. See Hightower (1972).
48. Center for Rural Affairs (1982:35).
49. The rate at which Iowa farmers left the land during the farm crisis is a matter of some controversy. The *total number* of Iowa farms declined by only about 1.4% a year during the farm crisis of the 1980s—the smallest percentage decline since the 1940s (*Des Moines Register* [1988c]). But subtracting the number of new farms that started operations during those same years makes a different picture emerge. By this yardstick, the situation for individual farmers during the farm crisis was indeed dire, with the percentage of farmers going out of business increasing dramatically, being at 1.9% in 1981, nearly doubling to 3.6% in 1984, and climbing to 5.5% in 1987 (Prairiefire [1988a:17]).

50. U.S. Congress. Office of Technology Assessment (1986:9).

51. Geisler and Popper (1984:8).

52. Geisler and Popper (1984:25).

53. Benjamin et al. (1986:3).

54. Land Stewardship Project (1987).

55. *Wall Street Journal* (1987).

56. Wessel (1983:118).

57. *Marshalltown [Iowa] Times-Republican* (1988:14).

58. For information on African-American farmers see *Catholic Rural Life* (1987), Brent (1987), Davidson (1987), and U.S. Department of Agriculture (1986).

59. *Des Moines Register* (1986e).

60. *Des Moines Register* (1988f).

61. *Des Moines Register* (1986c).

62. Wilson (1987:34).

63. President's National Advisory Commission on Rural Poverty (1967:ix).

64. For more on this period see Nelson (1979).

65. U.S. Department of Agriculture (1986:17).

66. U.S. Department of Agriculture (1986:17).

67. Strange (1988:117).

68. An acre-foot is the amount of water needed to cover one acre, one foot deep—or 325,851 gallons.

69. U.S. Congress. Joint Economic Committee (1986:225).

70. Worster (1985:313).

71. Sampson (1981:157).

72. Strange (1988:118).

73. U.S. Water Resources Council (1978:18).

74. Brown and Wolf (1984:8).

75. Lovins et al. (1984:69).

76. Paddock et al. (1986:7).

77. Lovins et al. (1984:81).

78. The Conservation Foundation (1986:18).

79. The Conservation Foundation (1986:23).

80. Brown and Wolf (1984:24).

81. Flaherty (1988:103).

82. *Des Moines Register* (1987i).

83. Postel (1988:119).

84. *U.S. News & World Report* (1987:70).

85. Williams (1987:49).
86. Freshwater Foundation (1986:4).
87. *Science* (1986:1491).
88. Strange (1988:203).
89. Meyerhoff and Mott (1985:30).
90. Burmeister et al. (1983).
91. *Hartford Courant* (1987).
92. Information on "cancer clusters" in California came from the following: *Los Angeles Times* (1985), *San Francisco Examiner* (1985), *Fresno Bee* (1985a, 1985b).
93. U.S. Department of Agriculture (1981:133).
94. Lillesand et al. (1977).
95. Hintz (1981:24).
96. Wiles (1985:306).

Chapter 3
The Rise of the Rural Ghetto

1. Padgett (1989) and Sheets (1989).
2. *Des Moines Register* (1988p).
3. In 1950, labor accounted for 38% of total farm inputs, while purchased inputs accounted for 45% of the total. By 1985, labor's share had dropped to 13%, while purchased inputs claimed 62% of total farm inputs. Leistritz and Murdock (1988:25).
4. Leistritz and Murdock (1988:25).
5. Leistritz and Murdock (1988:15).
6. Bender, Green, and Campbell (1971).
7. *Des Moines Register* (1986g). The problem is felt nationally; tractor sales fell 39% from 1979 to 1985 (*Des Moines Register* [1986a]).
8. *Des Moines Register* (1987l).
9. *Des Moines Register* (1987a).
10. *Des Moines Register* (1986f).
11. *Iowa City Press Citizen* (1987).
12. *Des Moines Register* (1986d).
13. As in other areas related to farming, agricultural banking is experiencing a seemingly inexorable trend toward concentration of ownership. A study commissioned by the American Bankers Association concluded that the largest banks will command a greater portion of agricultural financing in the future as many traditional rural banks consolidate or simply fade away (Leatham and Hopkin [1988]).

14. Swenson (1988:11–13).

15. *Des Moines Register* (1987c).

16. Swenson (1988).

17. Kusnet (1988:19).

18. Kusnet (1988:19).

19. This statistic and those in the following paragraph are found in Swenson (1988).

20. O'Hare (1988:12).

21. Swenson (1988). The decline hit Iowa's rural communities the hardest. Eight rural Iowa counties lost over 10% of their population from 1980 to 1987. And of the nine counties that gained population during those years, seven of them were overwhelmingly urban.

    A state can lose population in two ways. First, by an actual migration in which those individuals unable to find a good job simply move out-of-state in pursuit of work. Iowa lost many residents to this trend. But a state can also decline because of a drop in the number of babies born to its citizens. This, too, has happened in Iowa. Every year since 1980 has seen a decline in the number of babies born to Iowa women, to the point where in 1987 there were fewer babies born in the state than in any year since 1912. Even more significantly, the state's birth rate (the number of live births per 1,000 people) has declined to its lowest level since state records have been kept.

22. *Des Moines Register* (1987n).

23. *Des Moines Register* (1987m).

24. Western Interstate Commission for Higher Education (1988).

25. The rate of school consolidations has not reflected the dwindling student population. While Iowa ranks twenty-ninth among the states in terms of total population, it is thirteenth in the number of school districts (*Des Moines Register* [1987e]).

26. *Des Moines Register* (1988aa).

27. *Des Moines Register* (1988s).

28. The Corporation for Enterprise Development (1988:69). Between 1969 and 1987, the average annual salary of public school teachers in Iowa declined 8.2%—a larger drop than in any other state (U.S. Department of Education [1988:73]).

29. The loss of a school is especially devastating when other community institutions, such as church, social, and business organizations, have already dried up, which is frequently the case in communities at the point of losing a school. In Mechanicsville, Iowa, for example, the size of the congregation at the Methodist church declined from 310 to 246 members between 1975 and 1985.

A dwindling population alone does not account for the decline in community life in farm towns. As farm operations have grown steadily larger over the years, the free time needed for social activities has diminished. In a study of Maryland dairy farmers who had increased the size of their herds, one farmer said, "I used to be really active in the community when I had half the dairy herd size I now have. I need this herd size to meet the expenses of the farm; but now there's not enough time to do other things" (Poole [1981:120]).

30. *Des Moines Register* (1987h).
31. *Des Moines Register* (1987h).
32. *Des Moines Register* (1987d).
33. *Des Moines Register* (1988m).
34. *Des Moines Register* (1987b).
35. Associated Press (1988).
36. Padfield (1980).
37. U.S. Congress. Joint Economic Committee (1986:461).
38. Shanley (1988:6).
39. This was done because health-care costs were believed to be much lower in rural areas. While advocates of rural hospitals do not dispute that many of their costs are lower than those for urban hospitals, they do contend that the differential has been greatly exaggerated (Tauke [1986:2]).
40. *Des Moines Register* (1988k).
41. *Des Moines Register* (1987k).
42. Jacobsen and Albertson (1986:11).
43. U.S. Congress. Senate (1986:3).
44. Goldschmidt (1978).

Chapter 4
Poverty and Social Disintegration

1. Campbell (1927).
2. Agee and Evans ([1939] 1980).
3. President's National Advisory Commission on Rural Poverty (1967:xi).
4. U.S. Department of Agriculture. Economic Research Service (1985a:1). The rural poverty rate was 33.2% in 1959, and fell to 15.4% in 1980.
5. U.S. Department of Agriculture. Economic Research Service (1985a:3).

6. In this and in following sections, the terms "metropolitan," "urban," and "city" are used interchangeably to describe areas having a population of at least 50,000 residents, and include adjacent suburban areas. "Rural" refers to areas not included in the above definition. The terms "inner city" and "central city" are used synonymously to describe cities minus their adjacent suburbs.

7. Porter (1989:3). The poverty rate in 1987 for rural areas was 16.9%; for urban areas, 12.5%; and for central-city areas, 18.6%.

8. Prairiefire (1988a:13).

9. Shotland (1986:17). It is important to note, however, that farm families are more likely to be impoverished than are most other rural groups. One-third of all farm families fell below the poverty line in 1986 (Prairiefire [1988a:13]).

10. Porter (1989:27). While three-quarters of the rural poor are white, an African-American who happens to live in a rural area is far more likely to be poor than one living in a city—33% of all blacks in central cities live below the poverty line, while for rural blacks that figure jumps to an astonishing 44%. In 1987, 15.6% of the rural elderly lived below the poverty line, compared to 14.3% of the central-city elderly (the difference is not statistically significant). The poverty rate for young children (under six) in all rural single-mother households is 70.7%. The situation is even bleaker for young African-American children living in rural female-headed families: that group has a poverty rate of 81.2%.

11. Duncan and Tickamyer (1989:2).

12. U.S. Congress. Joint Economic Committee (1986:147). In 1980 the unemployment rate for metropolitan areas was 7.0%; for nonmetropolitan areas the rate was 7.3%.

13. U.S. Congress. Joint Economic Committee. Subcommittee on Agriculture and Transportation (1986:138). The rural unemployment rate for 1988, when factoring in these two groups, is 10.1%. Making the same changes in the urban rate produces an unemployment rate of 7.9%—far lower than the rural figure (Shapiro [1989:xiii]).

   Rural residents suffer disproportionately in another way. Among the unemployment subgroupings, urban and rural levels are closest in terms of straightforward unemployment. Since this figure is the one currently used by the government in targeting aid to a region, rural people—who suffer more in the categories *not* tracked by the government—receive fewer federal dollars than their situation merits.

14. Shapiro (1989:ix).

15. Shapiro (1989:xv).

16. Shapiro (1989:xv).
17. Porter (1989:26).
18. Ross and Danzinger (1987:2).
19. Swenson (1988:123).
20. Shotland and Loonin (1987).
21. Shotland (1986).
22. Physician Task Force on Hunger in America (1987:13–14).
23. Physician Task Force on Hunger in America (1986).
24. Shotland and Loonin (1988:7).
25. Shotland and Loonin (1988:25).
26. Shotland (1986:78).
27. Shotland (1986:134).
28. Porter (1989:30). Of the rural poor who live in families, 61% are in two-parent families, compared to a figure of 41.7% for their central-city counterparts. Married couples with children are ineligible to participate in AFDC in half the states.
29. Shotland and Loonin (1988:95–117). While 11.8% of urban poor children participated in the summer feeding programs in 1986, only 5% of their rural counterparts participated. Nonprofit organizations were excluded from the summer feeding programs because of concerns over sponsor abuse. Rather than banning all nonprofit organizations, Shotland and Loonin advocate implementing restrictions on the size of participating organizations as a way to limit abuses while still feeding large numbers of rural children.
30. Coons (1987:1).
31. Coons (1987:2).
32. Wright (1988).
33. Coons (1987:13).
34. Harrington (1985:101).
35. A secondary but important result of these family breakups is the increased need for foster care—a need that is not being met. In 1983, the state of Iowa cared for an average of 3,000 children each month. By 1989, that number had jumped by a third—to nearly 4,000 children a month (*Des Moines Register* [1989d]).
36. Associated Press (1987).
37. *Des Moines Register* (1988z).
38. Jacobsen and Albertson (1986:12).
39. *Des Moines Register* (1988g).
40. U.S. Congress. Joint Economic Committee. Democratic Staff (1986).
41. Dillman and Hobbs (1982:414).

Chapter 5
The Dying of the Light

1. *New York Times* (1985).
2. *USA Today* (1985).
3. *Time* (1985).
4. *Des Moines Register* (1988t).
5. Christiansen (1986).
6. Newman (1988:8).
7. Bluestone and Harrison (1982).
8. For an in-depth study of the PATCO strike and its effect on the participants, see Newman (1988:144–173).
9. *Des Moines Register* (1986b).
10. National Mental Health Association (1988:16).
11. Iowa State Department of Human Services (1988).
12. *Des Moines Register* (1988v).
13. Prairiefire (1988a:24).
14. National Mental Health Association (1988:22).
15. National Mental Health Association (1988:18).

Chapter 6
The Growth of Hate Groups

1. The money for the loans never materialized, and in 1986 Elliot was convicted of theft, fraud, and conspiracy.
2. Quoted in *The Hammer* (1984:27).
3. *Primrose and Cattlemen's Gazette* (1983:19).
4. Van Pelt (1984:3).
5. Trillin (1985:109–110).
6. Zeskind (1987:12).
7. For background information on David Duke see *The Monitor* (1988:4) and (1989:4).
8. Levitas and Zeskind (1986:2).
9. For more on these groups see Flynn and Gerhardt (1989) and Coates (1987).
10. For more on Identity see Zeskind (1986).
11. Zeskind (1986:5).
12. *The Monitor* (1989:5).
13. King (1989:375).

14. King (1989:142).

15. Levitas (1988:9).

16. Prairiefire (1988b).

17. Carto (1982:205).

18. Center for Democratic Renewal (1989).

19. Zeskind (1986:6).

20. *American POPULIST* (1986:4).

21. *The Monitor* (1989:5).

22. *The Monitor* (1989:5).

23. Quoted in Harrington (1984:192).

24. Coates (1987:257).

25. Zeskind (1986:30).

26. Baum (1978:x).

27. *Des Moines Register* (1989a).

28. *The Harris Survey* (1986).

29. *The Harris Survey* (1985).

30. Wallace (1925:24).

31. Goodwyn (1978:327).

32. Lynd and Lynd (1929:479).

Chapter 7
The Second Wave

1. *Des Moines Register* (1988r).

2. *Des Moines Register* (1988r).

3. *Des Moines Register* (1988e).

4. *Time* (1987).

5. *Des Moines Register* (1988q).

6. *Des Moines Register* (1988a).

7. Smith (1987:48–50).

8. U.S. Department of Agriculture. Economic Research Service (1988:xiv).

9. Strange et al. (1989:105). The policy of abandonment reached its apogee when a pair of urban geographers at Rutgers University proposed that the nation write off the entire Great Plains region. According to an article in the *Wall Street Journal*, the two advocate that the federal government "start buying back great chunks of the Plains, replant the grass, reintroduce the bison—and turn out the lights . . ." (*Wall Street Journal* [1989]).

10. *Des Moines Register* (1989f).

11. Strange et al. (1989:107). Even tiny rural communities have adopted a variation of this strategy, offering everything from free land to low-interest loans to families who move to town. Tiny Rolfe, Iowa, population 750, made national news in 1987 when it placed advertisements in newspapers throughout the country extolling the benefits of small-town life and offering new residents $1,200 in cash, a free lot to build on, and a complimentary membership in the local golf course and swimming pool. Although some 70 people took the community up on its offer in the next two years, some in town have had second thoughts about the program. "In some cases, [the new residents] are coming with nothing," said the administrator of a nearby charity. "They're having to be put up someplace because they don't have a place to stay or they're in subsidized housing." *Des Moines Register* (1989g).

12. East Central Iowa Economic Development Council (1987).

13. Cobb (1982:7–8).

14. Cobb (1982:23).

15. Cobb (1982:30).

16. Quillen (1986).

17. Cobb (1987:21).

18. For an in-depth study of the Saturn recruitment and its effects on the community of Spring Hill, Tennessee, see Garber (1988).

19. Garber (1988:i).

20. Garber (1988:35).

21. Summers (1986:183).

22. Bluestone and Harrison (1982:87).

23. *Des Moines Register* (1988d).

24. *Des Moines Register* (1988d).

25. Bruns (1988:27).

26. Miller (1987:36).

27. *Des Moines Register* (1989b).

28. *Chicago Tribune* (1987).

29. *Providence Journal* (1986).

30. For more on children and homework see Landrigan (1989).

31. Boris (1985).

32. *Chicago Tribune* (1987:26).

33. *Federal Register* (1988:45717).

34. U.S. Congress. General Accounting Office (1988:40–41).

35. Bluestone and Harrison (1982:228).

36. Bluestone and Harrison (1982:227).

Chapter 8
What Future, What Hope?

1. According to Webb (1959:352), "In the arid country the homestead law was of little avail; thousands of these homesteads reverted to the government yearly, while others passed into the hands of large land-owners, speculators, and stockmen."
2. Worster (1985:98).
3. Harrington (1985:218).
4. Polanyi (1944:164).
5. Barlett (1986:291).
6. *Des Moines Register* (1988l).
7. U.S. Department of Agriculture. Economic Research Service (1985b).
8. There is good reason to believe that farming is even more important to rural communities. The USDA defined nonmetropolitan counties as those containing cities of up to 50,000 people. Using the Census Bureau's definition of rural counties (including towns of up to only 2,500) would cut the total number of counties considered by two-thirds, thereby increasing the percentage of farm-dependent counties *Center for Rural Affairs* (1988:5).
9. U.S. Congress. Office of Technology Assessment (1986:9).
10. Strange (1988:80).
11. U.S. Department of Agriculture. Economics, Statistics and Cooperative Service (1979:112).
12. *Des Moines Register* (1988i).
13. In point of fact, the 1987 U.S. Senate bail-out of the Farm Credit System cost taxpayers $4 billion—more than the Chrysler bail-out did.
14. An industry is considered concentrated when just four companies account for over 40% of total sales. It is considered highly concentrated when 80% of total sales are made by four firms. Bannock et al. (1977:89).
15. Lee (1989:7).
16. *Des Moines Register* (1989c).
17. *Des Moines Register* (1988j).
18. *Des Moines Register* (1988x).
19. *Des Moines Register* (1988o). Even the successful fight against a buy-out can cost consumers dearly, as companies are forced to squander resources on resisting the takeover. When the retail food chain Safeway was forced to sell off its Los Angeles stores and acquire the

Lucky supermarket chain as part of an effort to fend off a takeover attempt, consumers paid for the costly moves through an estimated $357 million annual increase in food prices.

20. *Des Moines Register* (1987j).
21. *New York Times* (1986).
22. Chile's GNP in 1984 was $20.3 billion; Ecuador's GNP in 1985 was $12.1 billion (World Almanac [1987]).
23. A partial list of Cargill products includes:

Aluminum	Gasoline
Animal health products	Gold
Barge freight	Hemp
Barges	Hides
Barley	Hybrid seeds
Beef cattle	Insurance
Beef packing	Jute
Broilers	Lead
Chemicals	Linseed oil
Chocolate liquor	Liquid handling
Citrus pulp	Malt
Coffee	Manganese
Commodity futures	Manioc
Cooking oil	Molasses
Copper	Naphtha
Corn	Nickel
Corn gluten	Nitrogen
Corn meal	Oats
Corn starches	Ocean freight
Corn syrup	Orange juice
Cotton	Palm oil
Cottonseed	Peanut butter
Cottonseed meal	Peanut oil
Eggs	Pellet binders
Electronic parts	Pet foods
Feeds	Pig iron
Feedlots	Platinum
Feed supplements	Poultry
Fertilizers	Protein
Fibers	Rail freight
Financial instruments	Rapeseed
Flax	Resins
Flour	Rice
Foreign exchange	Rubber
Fuel oil	Rye

Salt	Sunflower hulls
Scrap metal	Sunflower meal
Seeds	Sunflower oil
Silver	Sunflower seeds
Sisal	Tallow
Sorghum	Tapioca
Soy beans	Tin
Soy flour	Turkeys
Soy meal	Waste management
Soy oil	Wheat
Starches	Wool
Steel	Zinc
Steel products	(Senate of Priests of St. Paul
Strapping material	and Minneapolis [1983:78–79]).
Sugar	

24. *Des Moines Register* (1988n).

25. *Des Moines Register* (1988x).

26. Leopold (1970:188).

27. National Research Council (1987).

28. See Weir and Schapiro (1981) and Norris (1982).

29. National Research Council (1989). Contrary to what some critics of the report have charged, the study does not advocate chemical-free farming It merely recommends the *prudent* use of chemicals, observing that farmers can often substitute less expensive and less environmentally hazardous methods of production.

30. *Des Moines Register* (1989e).

31. U.S. Department of Agriculture. Economic Research Service (1985b).

32. U.S. Congress. Office of Technology Assessment (1986:226).

33. See Heilbroner (1989). In Heilbroner (1966:72), the author also points out that "privilege under capitalism is much less 'visible,' especially to the favored groups, than privilege under other systems."

34. Lappe (1985:41).

35. U.S. Congress. Joint Economic Committee (1986:87).

36. Marty Strange, codirector of the Center for Rural Affairs, is the leading proponent of this system. Under Strange's plan, the government does not pay subsidies directly to farmers. Instead, grain purchasers are required to pay a "premium" to the farmer, for which they are later reimbursed by the government. The total amount of grain for which the government pays a premium is determined by adding together the anticipated domestic consumption of major crops, a reserve in case of crop failure, food, aid, and

export crops under contract. The plan is environmentally and economically far preferable to the present program. Since farmers would
receive a subsidy, or premium, for only a specific quantity of grain,
the government would not encourage maximum production—and
maximum environmental despoliation. Farmers would make production decisions based on the market, rather than on market-distorting
programs. Those interested in the plan should read the full description in Strange (1988:264–263).

37. It is because this kind of economic development is based on an antiquated way of thinking that so many of its products (homework, anti-
unionism) are throwbacks to that earlier era. As Meter (1987) points
out, "modern" industrialized farming is itself simply a reincarnation
of the old plantation system.

38. *Des Moines Register* (1988h).

39. Rohatyn (1981:20).

Coda: 1996

1. U.S. Department of Agriculture. Economic Research Service. 1995. *Rural Conditions and Trends.* Washington, D.C.: U.S. Government Printing Office, Spring, 14.

2. *Governing.* 1996. May, 18–27.

3. *Governing.* 1996. May, 18.

4. U.S. Department of Agriculture. Economic Research Service. 1995. *Rural Conditions and Trends.* Washington, D.C.: U.S. Government Printing Office, Spring, 18, 26. A living wage is defined as one sufficient to
bring a family of four above the poverty line, when employment is full-
time, year-round.

5. Telephone interview, 3 May 1996.

6. U.S. Department of Agriculture. Economic Research Service. 1995. *Rural Conditions and Trends.* Washington, D.C.: U.S. Government Printing Office, Spring, 5.

7. *Raleigh News and Observer*, 19 February 1995.

8. U.S. Department of Commerce. Bureau of the Census. *1992 Census of
Agriculture* [CD-ROM]. Table 15. Washington, D.C.: U.S. Government
Printing Office.

9. *Raleigh News and Observer*, 21 February 1995.

10. *False Patriots: The Threat of Antigovernment Extremists.* 1996. Montgomery, AL: Southern Poverty Law Center, 4.

Bibliography

ABRAMS, CHARLES. 1939. *Revolution in Land*. New York: Harper & Brothers.

AGEE, JAMES, AND WALKER EVANS. [1939] 1980. *Let Us Now Praise Famous Men*. Reprint. Boston: Houghton Mifflin Company.

The American POPULIST. 1986. June.

ASSOCIATED PRESS. 1987. Wire story, 16 July.

ASSOCIATED PRESS. 1988. "Aging Nebraska." Wire story, 18 August.

BANNOCK, GRAHAM, et al. 1977. *The Penguin Dictionary of Economics*. New York: Penguin Books

BARLETT, PEGGY. 1986. "Part-time Farming: Saving the Farm or Saving the Life-Style?" *Rural Sociologist*, Fall.

BARNES, PETER. 1975. *The People's Land*. Emmaus, PA: Rodale Press Book Division.

BAUM, GREGORY. 1978. Introduction to Charlotte Klein, *Anti-Judaism in Christian Theology*. Philadelphia: Fortress Press.

BEARD, CHARLES A. [1913] 1986. *An Economic Interpretation of the Constitution of the United States*. Reprint. New York: The Free Press.

BELDEN, JOSEPH N. 1986. *Dirt Rich, Dirt Poor*. New York and London: Routledge & Kegan Paul.

BENDER, L., B. GREEN, AND R. CAMPBELL. 1971. "The Process of Rural Poverty Ghettoization: Population *and* Poverty Growth in Rural Regions." Paper presented to the American Association for the Advancement of Science, Philadelphia, 28 December.

BENJAMIN, MEDIA, JOSEPH COLLINS, AND MICHAEL SCOTT. 1986. *No Free Lunch*. New York: Grove Press.

BENSON, C. RANDOLPH. 1971. *Thomas Jefferson as Social Scientist*. Rutherford, NJ: Fairleigh Dickinson University Press.

BENSON, EZRA TAFT. 1956. *Farmers at the Crossroads.* New York: The Devin-Adair Co.

BERRY, W. 1987. "A Defense of the Family Farm." In *Is There a Moral Obligation to Save the Family Farm?* Edited by G. Comstock. Ames: Iowa State University Press.

BLUESTONE, BARRY, AND BENNETT HARRISON. 1982. *The Deindustrialization of America.* New York: Basic Books.

BORIS, EILEEN. 1985. "Regulating Industrial Homework: The Triumph of 'Sacred Motherhood.' " *Journal of American History,* March, 745–763.

BRENT, PETE. 1987. "Black Farmers Face Extinction by Mid-90s." *Iowa Farm Unity News,* May, 2.

BROWN, LESTER, AND EDWARD WOLF. 1984. *Soil Erosion: Quiet Crisis in the World Economy.* Worldwatch Paper 60. Washington, DC: Worldwatch Institute.

BRUNS, MARY. 1988. "Landing Kodak's First Biotechnology Plant: Cedar Rapids, Iowa, as Development Center for Biotechnology." Paper prepared for Department of Urban and Regional Planning, University of Iowa, Spring.

BURMEISTER, LEON, et al. 1983. "Selected Cancer Mortality and Farm Practices in Iowa." *American Journal of Epidemiology,* Vol. 118, 1.

CAMPBELL, MACY. 1927. *Rural Life at the Crossroads.* Boston: Ginn and Company.

CARSON, RACHEL. 1962. *Silent Spring.* Boston: Houghton Mifflin Company.

CARTO, WILLIS. 1982. *Profiles in Populism.* Old Greenwich, CT: Flag Press.

Catholic Rural Life. 1987. July.

CENTER FOR DEMOCRATIC RENEWAL. 1989. *Ballot Box Bigotry: David Duke and the Populist Party.* Atlanta: Center for Democratic Renewal.

CENTER FOR RURAL AFFAIRS. 1982. *The Path Not Taken.* Walthill, NB: Center for Rural Affairs.

Center for Rural Affairs. 1988. Newsletter, November. Walthill, NB: Center for Rural Affairs.

Chicago Tribune. 1987. 8 March.

CHRISTIANSEN, SCOT. 1986. "The Rural Health Crisis." *Vital Signs,* University of Iowa Hospitals and Clinics, 7 November, 6.

COATES, JAMES. 1987. *Armed and Dangerous.* New York: Noonday Press.

COBB, JAMES C. 1982. *The Selling of the South.* Baton Rouge and London: Louisiana State University Press.

COBB, JAMES C. 1987. "Y'All Come Down!" *Southern Exposure*, no. 5–6.

CONAGRA. 1985. Annual Report.

THE CONSERVATION FOUNDATION. 1986. *Agriculture and the Environment in a Changing World Economy*. Washington, DC: The Conservation Foundation.

COONS, CHRISTOPHER. 1987. *Out in the Cold: Homelessness in Iowa*. Washington, DC: National Coalition for the Homeless.

THE CORPORATION FOR ENTERPRISE DEVELOPMENT. 1988. *Making the Grade: The 1988 Development Report Card for the States*. Washington, DC: The Corporation for Enterprise Development.

CREVECOEUR, J. HECTOR ST. JOHN DE. [1782] 1957. *Letters from an American Farmer*. Reprint. New York: E. P. Dutton.

DAVIDSON, OSHA. 1987. "Farms without Farmers." *The Progressive*, August, 25–27.

Des Moines Register. 1986a. 19 January, 5X.

Des Moines Register. 1986b. 16 February, 7A.

Des Moines Register. 1986c. 25 May, 3C.

Des Moines Register. 1986d. 10 July, 8S.

Des Moines Register. 1986e. 7 August, 2C.

Des Moines Register. 1986f. 23 November, 5F.

Des Moines Register. 1986g. 10 December, 1F.

Des Moines Register. 1986h. 19 December, 1A.

Des Moines Register. 1987a. 7 January, 1A.

Des Moines Register. 1987b. 10 January 1A.

Des Moines Register. 1987c. 25 January, 1F.

Des Moines Register. 1987d. 13 February, 1A.

Des Moines Register. 1987e. 22 February, 1A.

Des Moines Register. 1987f. 28 February.

Des Moines Register. 1987g. 8 March, 1F.

Des Moines Register. 1987h. 1 April, 3A.

Des Moines Register. 1987i. 26 October, 6A.

Des Moines Register. 1987j. 30 November, 4T.

Des Moines Register. 1987k. 1 December, 2A.

Des Moines Register. 1987l. 9 December, 2A.

Des Moines Register. 1987m. 26 December, 1A.

Des Moines Register. 1987n. 30 December, 1A.

Des Moines Register. 1988a. 6 January, 5T.

Des Moines Register. 1988b. 16 January, 7A.

Des Moines Register. 1988c. 8 February, 1A.

Des Moines Register. 1988d. 26 February, 7S.

Des Moines Register. 1988e. 18 March, 1A.

Des Moines Register. 1988f. 1 April, 7A.

Des Moines Register. 1988g. 4 April, 1A.

Des Moines Register. 1988h. 17 April, 1C.

Des Moines Register. 1988i. 24 April, 6A.

Des Moines Register. 1988j. 12 June, 3J.

Des Moines Register. 1988k. 14 June, 4A.

Des Moines Register. 1988l. 3 August.

Des Moines Register. 1988m. 9 August, 2A.

Des Moines Register. 1988n. 22 August, 4A.

Des Moines Register. 1988o. 28 August, 1W.

Des Moines Register. 1988p. 18 September, 2G.

Des Moines Register. 1988q. 20 September, 1A.

Des Moines Register. 1988r. 22 September, 1A.

Des Moines Register. 1988s. 28 September, 3A.

Des Moines Register. 1988t. 30 September, 1A.

Des Moines Register. 1988u. 1 October, 1S.

Des Moines Register. 1988v. 5 October, 3A.

Des Moines Register. 1988w. 11 October, 1S.

Des Moines Register. 1988x. 24 October, 1A.

Des Moines Register. 1988y. 7 November, 1A.

Des Moines Register. 1988z. 4 December, 1A.

Des Moines Register. 1988aa. 23 December, 5A.

Des Moines Register. 1989a. 6 January, 1A.

Des Moines Register. 1989b. 29 January, 1A.

Des Moines Register. 1989c. 5 February, 1W.

Des Moines Register. 1989d. 10 April, 1A.

Des Moines Register. 1989e. 8 September, 1A.

Des Moines Register. 1989f. 10 December, 1G.

Des Moines Register. 1989g. 13 December, 1A.

Des Moines Register. 1989h. 27 December, 1A.

DILLMAN, DON, AND DARYL HOBBS. 1982. *Rural Society in the U.S.: Issues for the 1980s.* Boulder, CO: Westview Press.

DUNCAN, CYNTHIA, AND ANN TICKAMYER. 1989. "The Rural Poor: What We Know and What We Need to Know." *Northwest Report,* March. Newsletter of the Northwest Area Foundation. St. Paul, MN.

EAST CENTRAL IOWA ECONOMIC DEVELOPMENT COUNCIL. 1987. *Economic Development Coordination Plan for Iowa's Merged Area X.* Cedar Rapids, IA, August.

ERIKSON, KAI. 1976. *Everything in Its Path.* New York: Simon and Schuster.

Federal Register. 1988. Vol. 53, no. 218, 10 November.

FLAHERTY, DIANE. 1988. "The Farm Crisis." In *The Imperiled Economy, Book II: Through the Safety Net.* Edited by Robert Cherry, Christine D'Onofrio, Cigdem Kurdas, Thomas Michl, Fred Moseley, and Michele Naples. New York: The Union for Radical Political Economics.

FLYNN, KEVIN, AND GARY GERHARDT. 1989. *The Silent Brotherhood: Inside America's Racist Underground.* New York: The Free Press.

FORD, ARTHUR M. 1973. *Political Economics of Rural Poverty in the South.* Cambridge, MA: Ballinger Publishing Company.

FRESHWATER FOUNDATION. 1986. *Pesticides and Groundwater: A Health Concern for the Midwest.* Navarre, MN: Freshwater Foundation.

Fresno Bee. 1985a. 17 July, B1

Fresno Bee. 1985b. 29 July, A1.

GARBER, CARTER. 1988. *Saturn: Tomorrow's Jobs, Yesterday's Wages and Myths.* Washington, DC: Rural Coalition, April.

GEISLER, CHARLES C. 1987. "Insights and Oversights: Selective Episodes in U.S. Land Reform." Paper presented at the Theology of Land Conference, Saint John's University, Collegeville, MN, 5 August.

GEISLER, CHARLES C., AND FRANK J. POPPER. 1984. *Land Reform, American Style.* Totowa, NJ: Rowman & Allanheld.

GELLHORN, MARTHA. 1936 *The Troubles I've Seen.* New York: Morrow.

GOLDSCHMIDT, WALTER. [1947] 1978. *As You Sow: Three Studies in the Social Consequences of Agribusiness.* Reprint. Montclair, NJ: Allanheld, Osmun & Co.

GOODWYN, LAWRENCE. 1978. *The Populist Moment.* New York: Oxford University Press.

The Hammer. 1984. "*Cattlemen's Gazette* Promotes Anti-Semitism." Spring, 25–28.

HARRINGTON, MICHAEL. [1984] 1985. *The New American Poverty.* Reprint. New York: Viking Penguin.

The Harris Survey. 1985. "Claims of Rising Anti-Semitism False." 30 May.

The Harris Survey. 1986. "A Study of Anti-Semitism in Rural Iowa and Nebraska." February.

The Hartford Courant. 1987. 24 May, 24.

HEILBRONER, R. 1966. *The Limits of American Capitalism.* New York: Harper & Row.

HEILBRONER, R. 1989. "The Triumph of Capitalism." *The New Yorker,* 23 January, 98–109.

HIGHTOWER, JIM. 1972. *Hard Tomatoes, Hard Times.* Cambridge, MA: Schenkman Publishing.

HINTZ, JOY. 1981. *Poverty, Prejudice, Power, Politics: Migrants Speak about Their Lives.* Columbus, OH: Avonelle Associates.

Iowa City Press Citizen. 1987. 5 January, 1D.

IOWA STATE DEPARTMENT OF HUMAN SERVICES. 1988. *A Statistical Overview of Program and Service Delivery.* Des Moines, March.

IOWA STATE UNIVERSITY. COOPERATIVE EXTENSION SERVICE. 1981. "Mechanicsville Community Attitude Survey." Ames: Iowa State University.

JACOBSEN, MICHAEL, AND BONNIE ALBERTSON. 1986. "Social and Economic Change in Rural Iowa: The Development of Rural Ghettos." Paper presented at the Eleventh National Institute on Social Work in Rural Areas, Harrisonburg, VA, July.

JEFFERSON, THOMAS. [1861] 1964. *Notes on the State of Virginia.* Reprint. New York: Harper & Row.

KAPLAN, SHEILA. 1987. "The Food Chain Gang." *Common Cause Magazine,* September-October, 12–15.

KING, DENNIS. 1989. *Lyndon LaRouche and the New American Fascism.* New York: Doubleday.

KREBS, A. V. 1988. "Corporate Agribusiness: Seeking Colonial Status for U.S. Farmers." *Multinational Monitor,* July-August, 19–21.

KUSNET, DAVID. 1988. "Where Have All the Good Jobs Gone?" *Multinational Monitor,* May, 19–20.

LAND STEWARDSHIP PROJECT. 1987. *Major Insurance Company Holdings (1986).* Marine-on-St. Croix, MN: Land Stewardship Project.

LANDRIGAN, PHILIP. 1989. "The Hazards to Children of Industrial Homework." Testimony before the U.S. Department of Labor, New York, 29 March.

LAPPE, FRANCIS MOORE. 1985. "The Family Farm: Caught in the Contradictions of American Values." *Agriculture and Human Values,* Spring, 36–43.

LEATHAM, DAVID, AND JOHN HOPKIN. 1988. "Transition in Agriculture: A Strategic Assessment of Agriculture and Banking." *Agribusiness,* March, 157–195.

LEE, THEA. 1989. "Choose Your Poison: Competition or Concentration." *Dollars & Sense,* July-August, 6–8.

LEISTRITZ, F. L., AND STEVE MURDOCK. 1988. "Financial Characteristics of Farms and of Farm Financial Markets and Policies in the United States." In *The Farm Financial Crisis.* Edited by S. Murdock and F. Larry Leistritz. Boulder, CO: Westview Press.

LEOPOLD, ALDO. [1949] 1970. *A Sand County Almanac.* Reprint. New York: Ballantine Books.

LEVITAS, DANIEL. 1988. "Food for Peace: Lyndon LaRouche's Latest Farmbelt Disinformation Campaign." Background report prepared for the American Jewish Committee. December.

LEVITAS, DANIEL, AND LEONARD ZESKIND. 1986. "The Farm Crisis and the Radical Right." Paper presented at the annual meeting of the Rural Sociologist Society, Salt Lake City, UT, 30 August.

LILLESAND et al. 1977. *An Estimate of the Number of Migrants and Seasonal Farmworkers in the U.S. and Puerto Rico.* Washington, DC: Legal Services Corporation.

LINGEMAN, RICHARD. 1980. *Small Town America.* Boston: Houghton Mifflin Company.

Los Angeles Times. 1985. 19 September, 1.

LOVINS, AMORY, L. HUNTER, AND MARTY BENDER. 1984. "Energy and Agriculture." In *Meeting the Expectations of the Land.* Edited by W. Jackson, W. Berry, and B. Colman. San Francisco: North Point Press.

LYND, ROBERT, AND HELEN LYND. 1929. *Middletown: A Study in Modern American Culture.* New York: Harcourt, Brace & World.

MADISON, JOHN. 1982. *Where the Sky Began.* San Francisco: Sierra Club Books.

Marshalltown [Iowa] *Times-Republican.* 1988. 19 April.

METER, KEN. 1987. "Why Save the Family Farm." *Seeds,* February, 6–13.

MEYERHOFF, ALBERT, AND LAWRIE MOTT. 1985. "Another Man's Poison." *The Amicus Journal,* Fall.

MILLER, MARC. 1987. "The Low Down on High Tech." *Southern Exposure,* no. 5–6.

MOBERG, DAVID. 1988. "Should We Save the Family Farm?" *Dissent,* Spring, 201–211.

The Monitor. 1988. "Populists Run Duke for President." Center for Democratic Renewal. November, 1–5.

The Monitor. 1989. "David Duke's Louisiana Victory Is Reflection of a National Strategy." May, 4–7.

MORGAN, DAN. 1979. *Merchants of Grain.* New York: The Viking Press.

MYRDAL, GUNNER. 1944. *An American Dilemma.* New York: Harper & Row.

NATIONAL MENTAL HEALTH ASSOCIATION. 1988. *Report of the National Action Commission on the Mental Health of Rural Americans.* Alexandria, VA: National Mental Health Association.

NATIONAL RESEARCH COUNCIL. 1987. *Regulating Pesticides in Food: The Delaney Paradox.* Washington, DC: National Academy Press.

NATIONAL RESEARCH COUNCIL. 1989. *Alternative Agriculture.* Washington, DC: National Academy Press.

NELSON, WILLIAM. 1979. "Black Rural Land Decline and Political Power." In *The Black Rural Landowner: Endangered Species.* Edited by Leo McGee and Robert Boone. Westport, CT: Greenwood Press, 83–96.

New York Times. 1985. 11 December, 1A.

New York Times. 1986. 30 March, 1B.

NEWMAN, KATHERINE. 1988. *Falling from Grace: The Experience of Downward Mobility in the American Middle Class.* New York: The Free Press.

NORRIS, RUTH, ED. 1982. *Pills, Pesticides and Profits.* Croton-on-Hudson, New York: North River Press.

NYE, RUSSEL. 1961. "Has the Midwest Ceased to Protest?" In *The Midwest: Myth or Reality.* Edited by Thomas McAvoy. Notre Dame, IN: University of Notre Dame Press.

O'HARE, WILLIAM. 1988. *The Rise of Poverty in Rural America.* Population Trends and Public Policy. Report no. 15. Washington, DC: Population Reference Bureau, July.

OPIE, JOHN. 1987. *The Law of the Land.* Lincoln: University of Nebraska Press.

PADDOCK, JOE, NANCY PADDOCK, AND CAROL BLY. 1986. *Soil and Survival.* San Francisco: Sierra Club Books.

PADFIELD, HARLAND. 1980. "The Expendable Rural Community and the Denial of Powerlessness." In *The Dying Community.* Edited by Art Gallaher, Jr., and Harland Padfield. Albuquerque, NM: University of New Mexico Press.

PADGETT, TIM. 1989. "Just Saying No to Wal-Mart." *Newsweek,* 13 November, 65.

PHYSICIAN TASK FORCE ON HUNGER IN AMERICA. 1986. *Hunger Counties 1986: The Distribution of America's High Risk Areas.* Boston: Harvard School of Public Health, January.

PHYSICIAN TASK FORCE ON HUNGER IN AMERICA. 1987. *Hunger Reaches Blue Collar America.* Boston: Harvard School of Public Health, October.

PIVEN, FRANCES FOX, AND RICHARD CLOWARD. 1972. *Regulating the Poor: The Functions of Public Welfare.* New York: Vintage Books.

POLANYI, KARL. 1944. *The Great Transformation.* New York: Farrar and Rinehart.

POOLE, DENNIS. 1981. "Farm Scale, Family Life, and Community Participation." *The Journal of Rural Sociology,* Spring.

PORTER, KATHRYN. 1989. *Poverty in Rural America: A National Overview.* Washington, DC: Center on Budget and Policy Priorities.

POSTEL, SANDRA. 1988. "Controlling Toxic Chemicals." In *State of the World, 1988.* Edited by Lester Brown. New York: W. W. Norton & Company.

PRAIRIEFIRE. 1988a. *No Place to Be: Farm and Rural Poverty in America.* Des Moines: Prairiefire Rural Action, December.

PRAIRIEFIRE. 1988b. Confidential memorandum. 15 August.

PRESIDENT'S NATIONAL ADVISORY COMMISSION ON RURAL POVERTY. 1967. *The People Left Behind.* Washington, DC: U.S. Government Printing Office, September.

The Primrose and Cattlemen's Gazette. 1983. 7 June.

Providence Journal. 1986. 21 August.

QUILLEN, ED. 1986. "Soft Path in the Rockies: At Home with Hunter and Amory Lovins." *Blair and Ketchum's Country Journal,* July.

ROHATYN, FELIX. 1981. "Reconstructing America." *New York Review of Books,* 5 March, 16–20.

ROSS, CHRISTINE, AND SHELDON DANZINGER. 1987. "Poverty Rates by State, 1979 and 1985: A Research Note." *Focus,* University of Wisconsin Madison, Institute for Research on Poverty. Fall, 1–5.

SAGE, LELAND. 1974. *A History of Iowa.* Ames: The Iowa State University Press.

SAMPSON, R. NEIL. 1981. *Farmland or Wasteland.* Emmaus, PA: Rodale Press.

San Francisco Examiner. 1985. 21 July.

Science. 1986. "Groundwater Ills: Many Diagnoses, Few Remedies." 20 June, 1491.

SENATE OF PRIESTS OF ST. PAUL AND MINNEAPOLIS. 1983. *Daily Bread: An Abdication of Power.* St. Paul and Minneapolis: The Priests' Senate.

SHANLEY, MARY KAY. 1988. "The Economic Struggle for Rural Health Care." *Iowa Commerce,* March-April, 6–9.

SHAPIRO, ISAAC. 1989. *Laboring for Less.* Washington, DC: Center on Budget and Policy Priorities, October.

SHEETS, KENNETH. 1989. "How Wal-Mart Hits Main St." *U.S. News & World Report,* 13 March, 53–55.

SHOTLAND, JEFFREY. 1986. *Rising Poverty, Declining Health: The Nutritional Status of the Rural Poor.* Washington, DC: Public Voice for Food and Health Policy, February.

SHOTLAND, JEFFREY, AND DEANNE LOONIN. 1988. *Patterns of Risk: The Nutritional Status of the Rural Poor.* Washington, DC: Public Voice for Food and Health Policy, February.

SOLKOFF, JOEL. 1985. *The Politics of Food.* San Francisco: Sierra Club Books.

SMITH, HENRY NASH. 1950. *Virgin Land.* Cambridge and London: Harvard University Press.

SMITH, RUSSELL, ED. 1987. *Nebraska Policy Choices.* Omaha: University of Nebraska at Omaha.

STATEN, JAY. 1987. *The Embattled Farmer.* Golden, CO: Fulcrum.

STRANGE, MARTY. 1988. *Family Farming.* Lincoln: University of Nebraska Press.

STRANGE, MARTY, et al. 1989. *Half a Glass of Water: State Economic Development Policies and the Small Agricultural Communities of the Middle Border.* Walthill, NB: Center for Rural Affairs.

SUMMERS, GENE. 1986. "Rural Industrialization." *Rural Sociologist,* vol. 6, 3.

SWENSON, DAVID. 1988. *A Decade of Adjustment.* Iowa City: The University of Iowa. Institute for Public Affairs.

TAUKE, THOMAS. 1986. Statement made before the U.S. Senate Finance Committee, 9 May. Photocopy.

Time. 1985. 23 December, 26.

Time. 1987. 3 August.

TRILLIN, CALVIN. 1985. "I've Got Problems." *The New Yorker,* 18 March, 109–117.

USA Today. 1985. 11 December, 1.

U.S. CONGRESS. GENERAL ACCOUNTING OFFICE. 1988. *Enterprise Zones: Lessons from the Maryland Experience.* Washington, DC: U.S. Government Printing Office, December.

U.S. CONGRESS. HOUSE. SPECIAL COMMISSION ON FARM TENANCY. 1937. *Farm Tenancy.* 75th Cong., 1st sess. House Document no. 149. Washington, DC: U.S. Government Printing Office.

U.S. CONGRESS. JOINT ECONOMIC COMMITTEE. DEMOCRATIC STAFF. 1986. *The Bi-Coastal Economy.* Washington, DC: U.S. Government Printing Office, July.

U.S. CONGRESS. JOINT ECONOMIC COMMITTEE. SUBCOMMITTEE ON AGRICULTURE AND TRANSPORTATION. 1986. *New Dimensions in Rural Policy:*

Building upon Our Heritage. 99th Cong., 2nd sess. Washington, DC: U.S. Government Printing Office.

U.S. CONGRESS. OFFICE OF TECHNOLOGY ASSESSMENT. 1986. *Technology, Public Policy, and the Changing Structure of American Agriculture.* Washington, DC: U.S. Government Printing Office, March.

U.S. CONGRESS. SENATE. COMMITTEE ON GOVERNMENTAL AFFAIRS. SUBCOMMITTEE ON INTERGOVERNMENTAL RELATIONS. 1986. *Governing the Heartland: Can Rural Communities Survive the Farm Crisis?* 99th Cong., 2nd sess. Draft Committee Print. Washington, DC: U.S. Government Printing Office.

U.S. DEPARTMENT OF AGRICULTURE. 1981. *A Time to Choose: Summary Report on the Structure of Agriculture.* Washington, DC: U.S. Government Printing Office, January.

U.S. DEPARTMENT OF AGRICULTURE. ECONOMIC RESEARCH SERVICE. 1983. *Economic Indicators of the Farm Sector, State Income and Balance Sheet Statistics.* Washington, DC: U.S. Government Printing Office.

U.S. DEPARTMENT OF AGRICULTURE. ECONOMIC RESEARCH SERVICE. 1985a. *Characteristics of Poverty in Non-Metro Counties.* Prepared by Elizabeth Morrissey. Washington, DC: U.S. Government Printing Office, July.

U.S. DEPARTMENT OF AGRICULTURE. ECONOMIC RESEARCH SERVICE. 1985b. *The Diverse Social and Economic Structure of Nonmetropolitan American.* Prepared by Lloyd Bender et al. Rural Development Research Report no. 49. Washington, DC: U.S. Government Printing Office, September.

U.S. DEPARTMENT OF AGRICULTURE. ECONOMIC RESEARCH SERVICE. 1988. *Rural Economic Development in the 1980s: Prospects for the Future.* Rural Development Research Report no. 69. Washington, DC: U.S. Government Printing Office, September.

U.S. DEPARTMENT OF AGRICULTURE. ECONOMICS, STATISTICS AND COOPERATIVE SERVICE. 1979. *Structure Issues of American Agriculture.* Agricultural Economic Research Report no. 438. Washington, DC: U.S. Government Printing Office.

U.S. DEPARTMENT OF AGRICULTURE. ECONOMICS, STATISTICS AND COOPERATIVE SERVICE. 1986. *Black Farmers and Their Farms. Prepared by Vera Banks. Rural Development Research Report no. 59. Washington, DC: U.S. Government Printing Office.*

U.S. DEPARTMENT OF EDUCATION. NATIONAL CENTER FOR EDUCATION STATISTICS. 1988. *Digest of Education Statistics.* Washington, DC: U.S. Government Printing Office.

U.S. News & World Report. 1987. 16 November, 70.

U.S. WATER RESOURCES COUNCIL. 1978. *The Nation's Water Resources, 1975–2000*. Summary Report. Washington, DC: U.S. Government Printing Office. December.

VAN PELT, JUDGE SAMUEL. 1984. "Report to Governor Robert Kerry on the Death of Arthur L. Kirk." 1 December.

Wall Street Journal. 1987. 3 February, 1.

Wall Street Journal. 1989. 16 August, 1.

WALLACE, HENRY C. 1925. *Our Debt and Duty to the Farmer*. New York: The Century Co.

WEBB, WALTER PRESCOTT. 1931. *The Great Plains*. Boston: Ginn.

WEIR, DAVID, AND MARK SCHAPIRO. 1981. *Circle of Poison*. San Francisco: Food First Books.

WESSEL, JAMES. 1983. *Trading the Future*. San Francisco: Institute for Food and Development Policy.

WESTERN INTERSTATE COMMISSION FOR HIGHER EDUCATION. 1988. *High School Graduates: Projections by State, 1986 to 2004*. Boulder, CO.

WILES, RICHARD. 1985. "Pesticide Risk to Farm Workers." *The Nation*, 5 October, 306.

WILLIAMS, TED. 1987. " 'Silent Spring' Revisited." *Modern Maturity*, October–November.

WILSON, WILLIAM J. 1987. *The Truly Disadvantaged*. Chicago: The University of Chicago Press.

WORLD ALMANAC. 1987. *The World Almanac and Book of Facts 1988*. New York: Newspaper Enterprise Association.

WORSTER, DONALD. 1985. *Rivers of Empire: Water, Aridity and the Growth of the American West*. New York: Pantheon Books.

WRIGHT, R. DEAN. 1988. *Executive Summary of the Final Report Pertaining to the Problem of Homeless Children and Children of Homeless Families in Iowa*. Des Moines, IA: Drake University.

ZESKIND, LEONARD. 1986. *The "Christian Identity" Movement: A Theological Justification for Racist and Anti-Semitic Violence*. Atlanta: Center for Democratic Renewal, October.

ZESKIND, LEONARD. 1987. Introduction to Chris Lutz, *They Don't All Wear White Sheets*. Atlanta: Center for Democratic Renewal.

ZINN, HOWARD. 1980. *A People's History of the United States*. New York: Harper & Row.

Index

Index to Coda